**Modelling for Field Biologists
and Other Interesting People**

Students of evolutionary and behavioural ecology are often unfamiliar with
mathematical techniques, even though much of biology relies on mathematics.
Evolutionary ideas are often complex, meaning that the logic of hypotheses
proposed should be tested not only empirically but also mathematically. There are
numerous different modelling tools used by ecologists, ranging from
population genetic 'bookkeeping' to game theory and individual-based computer
simulations. Due to the many different modelling options available, it is often
difficult to know where to start. Hanna Kokko has designed this book to help
with these decisions. Each method described is illustrated with one or two biologically interesting examples that have been chosen to help to overcome the fears
of many biologists when faced with mathematical work, while also
providing the programming code (MATLAB®) for each problem. Aimed primarily
at students of evolutionary and behavioural ecology, this book will be of use to
any biologist interested in mathematical modelling.

HANNA KOKKO is a Professor in the Department of Biological and
Environmental Science at the University of Helsinki.

Modelling for Field Biologists and Other Interesting People

HANNA KOKKO

*Department of Biological and Environmental Science,
University of Helsinki, Finland*

CAMBRIDGE
UNIVERSITY PRESS

University Printing House, Cambridge CB2 8BS, United Kingdom

One Liberty Plaza, 20th Floor, New York, NY 10006, USA

477 Williamstown Road, Port Melbourne, VIC 3207, Australia

314-321, 3rd Floor, Plot 3, Splendor Forum, Jasola District Centre, New Delhi - 110025, India

103 Penang Road, #05-06/07, Visioncrest Commercial, Singapore 238467

Cambridge University Press is part of the University of Cambridge.

It furthers the University's mission by disseminating knowledge in the pursuit of education, learning and research at the highest international levels of excellence.

www.cambridge.org
Information on this title: www.cambridge.org/9780521831321

First published 2007
Reprinted with corrections 2008

A catalogue record for this publication is available from the British Library

ISBN 978-0-521-83132-1 Hardback
ISBN 978-0-521-53856-5 Paperback

To my parents,
who gave me all those Moomin books to read,
and to the memory of Tove Jansson,
who wrote them

Contents

Preface

This book arose from far too many queries addressed to me in the past 10 years by behavioural or evolutionary ecologists: 'If I wanted to learn to model, what book should I read?' I could not give a straightforward answer to this question, and neither could I choose a course book to cover modelling in behavioural or evolutionary ecology at a beginner's level for the courses I have been teaching. There are many books available that delve into particular methods, but grabbing them requires that one knows a priori that the particular method will be useful. I was looking for a book that would provide a gentle enough introduction for people who might range from keen undergraduates to experienced researchers, but share one thing in common: no hands-on experience with mathematical modelling so far. I found some very good texts for population ecologists, but nothing similar for evolutionary or behavioural ecologists.

This book is an attempt to fill in this gap. My intention is not to provide full coverage of all mathematical tools used in evolutionary studies today – that would be far too much to ask from a volume of this size. Instead, my intention is to present an entry-level 'toolbox' for those people who lack nothing but a kick-start to add modelling techniques to their repertoire of scientific skills. The subtitle 'and other interesting people' reflects the attitude that I have tried to follow in this book: there is no need for anyone to be intimidated by modelling work, let alone by people who possess this apparently magical skill. Working with real-life questions can be so much more interesting – and adding some modelling skills can be much fun. Whether the reader is an undergraduate or a senior scientist, my intention is to show how one might approach a problem, and – if one particular method grabs their attention – give suggestions where to read on to learn more.

I have also been lucky to learn that there are now more books that are devoted to teaching modelling to biologists than there were 10 years ago, which makes my task of pointing to other sources vastly easier. My goal is not to supersede the superb introductions that now exist for many, perhaps most, of the methods presented here. Instead, my intention is to give a glimpse of what is available, through the use of examples. Each chapter illustrates the use of a particular method: say, dynamic optimization, or quantitative genetics. I present the simplest example that I could think of that is complex enough to interest people who probably already lead exciting lives studying the wonders of nature, or are simply keen to learn more.

Absolutely no prior knowledge of any of the methods is required, while familiarity with evolutionary thinking in general will be assumed. My style is intentionally informal and 'chatty'. This is to appreciate what I imagine a likely reader of my book to be like: an engaged and intelligent person, who has probably spent more time wearing muddy boots or rainforest-mouldy T-shirts than staring at books heavily laden with mathematical expressions – these being either too dry or too scary in his or her opinion. To keep to this style, I will have to disappoint those who look for a full treatment of the methods. In my experience, formal definitions of mathematical concepts are far easier to find in the literature than friendly, entry-level explanations of what they mean. For this reason, I am concentrating on providing the latter type of information, with a focus on illustrative examples, while also giving pointers to the mathematically more complete texts in which the full derivations are given. The aim of this book is, therefore, to fill in the gap between the would-be modeller and the beginning level of other, more complete and thorough texts available. Finally, I am hoping to provide some food for thought for those scientists who have some experience with building models but are not familiar with too many different techniques.

Modelling relies heavily on computers nowadays. The development of ideas in the book is not specific to any particular programming language, but to help readers who might be interested in the programming aspect. All the examples in the book are available electronically, with the programming code for reproducing all the figures at www.helsinki.fi/~hmkokko/modelling. For this I have used MATLAB (www.mathworks.com), as it is a versatile tool that works well for ecologists. For university students or staff, it is worth knowing that universities have licences for this program more often than biologists seem to realise. Nevertheless, not all readers of this book will have access to this particular program. They might also like to stick to another program for some other reason – or use the old-fashioned but often surprisingly useful pen and paper method. I strongly encourage such readers to read on: the programming language is a side issue here, used in technical side boxes only. It really does not matter if Hamilton's rule is expressed as if $b*r>c$ (MATLAB) or as $=IF$ $(A1*A2>A3;1;0)$ (Microsoft Excel, assuming that values of b, r and c are stored in cells A1, A2 and A3, respectively). Therefore the examples can all be translated across programming languages, although spreadsheet programs may become cumbersome to use for more extensive calculations. Two examples of freeware programs very similar to MATLAB are SysQuake LE (www.calerga.com), and the somewhat more statistically oriented R (www.r-project.org). A "free" version of MATLAB is Scilab (http://www.scilab.org/). Many prefer 'traditional' programming languages such as C or Basic (nowadays often in the version VisualBasic), which is also fine. Ideas are important, not the programming platform used. An open source platform called Scilab (www.scilab.org) is also available.

An additional reason why the programming language does not matter is that often the goal is to derive analytical expressions. These are general solutions such as $br > c$, as opposed to extensive lists of numerical values. This means that programming languages are not always even needed to find out expected evolutionary outcomes. Even here, the software industry has been busy producing equation-solving beasts such as MAPLE (www.maplesoft.com) or MATHEMATICA (www.wolfram.com), but the humble combination of pen and paper surprisingly often retains its centuries-old effectiveness. I still very often find the neatest presentation of an equation fastest by scribbling all those squiggles on paper, and few of the examples presented in this book are computing intensive at all.

Books are rarely created alone, and this one is no exception. My sincere thanks go to my Cambridge University Press editor Dominic Lewis, who allowed and encouraged me to write in a relaxed style that would make any editor of a scientific journal cringe. His help and support was absolutely crucial as was the editing work by Jane Ward. The other important support group is students. This book is heavily based on courses I have given at Jyväskylä University and Helsinki University in the years 2002–2006. My very warm thanks go to all the participants: it is incredible to see such active and enthusiastic students, and the feedback helped me more than you may believe in this project. Some of the feedback diaries you produced should have been called works of art, and I have ruthlessly exploited all the insights they contained. Andrés López-Sepulcre and Daniel Rankin helped me a lot with the courses – thank you.

A number of people read individual chapters with great care, some of them even commented on the whole manuscript in impressive detail. The feedback has been truly indispensable. Patricia Backwell, Mats Björklund, Anders Brodin, Rob Brooks, Jakob Bro-Jørgensen, Johanna Eklund, Kevin Foster, Nika Galic, Phillip Gienapp, Ilkka Hanski, Wade Hazel, Mikko Heino, Alasdair Houston, Michael Jennions, Jonathan Jeschke, Jussi Lehtonen, Anna Lindholm, Jan Lindström, Andrés López-Sepulcre, Martim Melo, Hans Metz, Lesley Morrell, Päivi Paavilainen, Janne Pyykkö, Esa Ranta, Ian Rickard, Walter Rydman, Franziska Schädelin, Toomas Tammaru, Andrea Townsend, Wouter Vahl, and a number of students whose views were transmitted to me via the above-mentioned people: thank you! Michael Jennions deserves a special mention for coming up with the title of the book, and Martim Melo for the quote from *Travels of Praiseworthy Men*. The book has immensely benefited from all these comments, but any errors that remain, as well as any less than perfect choices regarding style and content, are obviously mine.

Finally, thank you Liisa and Ilkka.

1

Modelling philosophy

where we get momentarily lost in a forest, but emerge intact

Figure 1.1 shows three different kinds of model. A supermodel like Naomi Campbell presents, to some of us at least, an 'idealized' concept of a human being (Fig. 1.1a). The miniature model shown in Fig. 1.1b was built by my uncle to show what his home town Kuopio looked like in the 1930s. Finally, Fig. 1.1c is a mathematical description of the dynamics of a two-species system of a predator and a prey species.

The models all look very different. They also differ a lot in how scary they look to the average behavioural or evolutionary ecologist – most will have to resist the temptation to close their eyes when encountering Fig. 1.1c, together with its equations, much more than when looking at Fig. 1.1a. My aim in this chapter is to show that there is indeed a reason why the word 'model' is used to describe all these figures, to rectify some common misconceptions about models (especially the mathematical ones), and to make life a little less scary for those who know they should be more familiar with modelling in behavioural or evolutionary ecology than they currently are.

Ecology is defined as a science that investigates the abundance and distribution of organisms. This may first sound a little boring, but it gets more interesting once one notices that interactions between organisms play a crucial role here. This means that ecology must study the causalities that underlie the changes in individual numbers, rather than merely providing us with simplistic methodologies of bookkeeping. When trying to understanding those causalities, evolutionary aspects must be taken into account, because evolution underlies everything that organisms do. Very often, the interactions manifest themselves in the behaviour of the

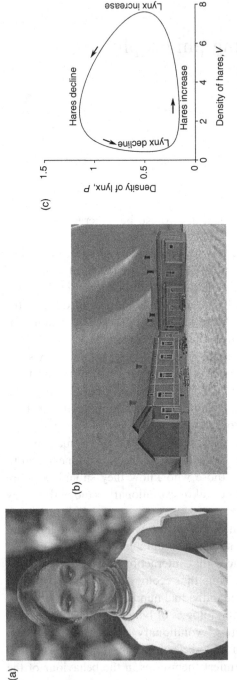

Fig. 1.1 Models. (a) Naomi Campbell; (b) a street corner from a miniature model of Kuopio, a town in Eastern Finland; (c) hare–lynx cycle described by Lotka–Volterra equations $dV/dt = rV - aVP$ and $dP/dt = -qP + baVP$. Here V is hare density; P is lynx density; t marks time such that dV/dt gives the rate of change in the hare density and dP/dt similarly for lynx; r, a, b and q are parameters that determine how quickly hares reproduce (r), how many hares are eaten (a) for a given density of lynx and hare, how efficiently eaten hares are converted into lynx offspring (b) and how quickly lynx die when food is in short supply (q). The limit cycle is drawn using values $r = 0.1$, $a = 0.2$, $b = 0.1$, and $q = 0.1$. The predator–prey cycle shows qualitatively nice cycling, but the model is meant to be conceptual only: parameters are hardly realistic when at the peak of lynx density there are only twice as many hares as there are lynx in the forest. The Lotka–Volterra cycle can be found in almost all textbooks on ecology. I have used the notation of Odenbaugh (2005), who uses the predator–prey cycle to illustrate several important philosophical issues on the need to simplify when modelling ecological phenomena.

organism in question. Behavioural and evolutionary ecology are the sciences that study the evolutionary and ecological causalities that cause individuals to behave the way they do – and while saying this, one should not forget the physiological and physical mechanisms that shape and constrain the ways in which individuals can behave.

This definition places behavioural and evolutionary ecology in a scarily complex web of interactions. It appears that everything interacts with everything else: a migrating bird has to combat weather and winds, manage its energy reserves, find the flight speed that is appropriate for the wing shape the bird has, find its way using perhaps several different orientation mechanisms, avoid predators on the way, find good stopover sites, arrive in a sensible time of the year and compete with conspecifics for breeding localities and possibly mates too. And, all of this is governed by genes that influence the bird's behaviour in a multitude of ways. Why should we ever be interested in modelling such a system – or even if we were, what hope do we have of ever capturing the complexity of the situation the bird finds itself in?

The quick answer is that there is no hope. Our model simply will never be able to deal with such complexity. The perhaps more surprising elaboration of this statement is that if a model did capture all of this – perhaps computers in the future could stomach it all? – the outcome would not be desirable at all. Why? The famed author J. A. Suarez Miranda said it all in his book *Travels of Praiseworthy Men*, already in 1658:

> ... In that Empire, the craft of Cartography attained such Perfection that the Map of a Single province covered the space of an entire City, and the Map of the Empire itself an entire Province. In the course of Time, these Extensive maps were found somehow wanting, and so the College of Cartographers evolved a Map of the Empire that was of the same Scale as the Empire and that coincided with it point for point. Less attentive to the Study of Cartography, succeeding Generations came to judge a map of such Magnitude cumbersome, and, not without Irreverence, they abandoned it to the Rigours of sun and Rain. In the western Deserts, tattered Fragments of the Map are still to be found, Sheltering an occasional Beast or beggar; in the whole Nation, no other relic is left of the Discipline of Geography.

If this passage makes you want to find the rest of Suarez Miranda's book in a library (instead of learning more about modelling), prepare to be disappointed. The book or its author never existed except in the above quote written by Jorge Luis Borges and Adolfo Bioy Casares (see Borges

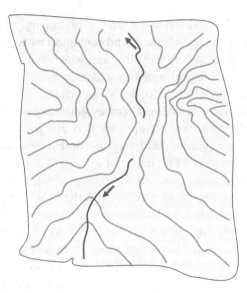

Fig. 1.2 A map is a model too. How would you plan your route in this
landscape?

1975). Nevertheless, it marks a perfect beginning for a tour of the virtual
world of modelling.

Why do we make maps of landscapes? These exist at a range of different scales, and the detail included may also vary irrespective of scale: a
mining company probably requires different information about an area
than a passing tourist or a hiker. Using the map in Fig. 1.2, a hiker is
probably quite keen to have a look at the contour lines: they tell her the
easiest route through a mountain pass and predict the direction of the
flow of the river she will encounter. This can be important for a variety of
reasons, including orientation purposes. Now, to predict the flow of the
river, it is not necessary to mark every tuft of grass that borders the river,
or every tree in the forest our hiker is passing through, on the map. These
would only make it more difficult to grasp the overall shape of the
landscape. Taken to the extreme, if the ultimate goal was to include
all the detail of the forest in the map, our poor hiker would have to carry
the whole landscape with her. Even if this was physically possible, the
gain from doing so would be zero: if lost in a forest, she would not
become any wiser by looking at a too vastly detailed map than by staring
at the original forest. Such maps indeed deserve to be left rotting in
the rain.

This, of course, is exactly the reason why models are 'not real'. They are idealized versions of the real world. My uncle is now building a street with shops in his miniature model of Kuopio. He knows that to create the visual impression one would get by looking down a particular street in Kuopio, it is important to shape the fonts in shop windows to match the original ones, and he has made countless visits to the local library's archives to verify all this detail. However, it is far less important to use exactly the same chemical composition of the paint. This is the art of modelling: to know which aspects of reality one can sacrifice, and which ones are crucial to retain. Any medical simulation of drug concentrations in various tissues of the human body is likely to include the fact that the blood vessels from the gut lead to the liver, since that organ rids us of strange compounds in the bloodstream. The whole machinery involved is never included. For example, the developmental fact that the DNA in of our cells contain instructions on how to build the blood vessels is simply not relevant – until, perhaps, we find patients who have a genetic disorder that disrupts this process and makes them react to medicines in surprising ways. But even then, we should not go to the other extreme and start including every possible gene action in the model, from eye colour determination to how fast one's toenails grow.

George Box, a well-known statistician, once quipped 'All models are false, but some models are useful'. Have a look again at Fig. 1.1c. Here, V might denote the numbers of hares in an area, and P are lynx. The equations may look scary, but all they do is to describe some numerical rules of species interaction: dV/dt is the change of hare numbers over time, and this is larger if there are lots of hares (large V) – since many hares can produce more leverets (baby hares) than few hares can. Likewise, lots of lynx can produce lots of new lynx. How lynx and hare populations respond to each other, however, differs between the species. Lynx populations tend to increase when hares are abundant, but hare populations respond negatively to the abundance of lynx, for rather obvious reasons. This is reflected in a negative sign in front of P in the hare equation, and in a positive sign in front of V in the lynx equation. The squiggles in Fig. 1.1c are called Lotka–Volterra equations, and they predict – when solved – that lynx and hare numbers will cycle up and down, hares always declining before their predators do.

Of course, the model is incomplete. It is wrong. It does not include density-dependent responses in the absence of predation: even if hares bred like, well, rabbits, neither species can in reality fill every square metre of the world with their offspring – food would run out long before that. Also, the model does not include the fact that both hares and lynx need some time before they mature. Or that both species reproduce sexually. Or that hares could possess antipredator responses. Or that lynx might use different hunting techniques in the summer and the winter. And so on ad infinitum. All these points are true, but the important point is: would one have predicted that the simple statements "hares beget hares; lynx beget lynx; and lynx eat hares" can, *by themselves*, lead to predator–prey cycles? Perhaps some people's intuition tells them immediately that this is the case. But perhaps someone else disagrees and argues that some other process, perhaps some cyclic weather pattern, must interfere before cycles can occur. How could we ever solve the case, without writing the matter down in equations? By carefully looking at patterns in nature? Not easy: we simply cannot find idealized hare and lynx populations that would not have interfering factors of all sorts. By conducting clever experiments? Yes, this definitely plays a large part: experiments very typically try to strip down the messy nature of reality to get at the pure effect of a particular biological factor. Nevertheless, there is always the chance that some other factor that we did not think about at all was influencing our results. In the end, the only way to find out if an argument 'from X follows Y' is a valid statement to offer, one has to put it in a mathematical form.

The analogy with an experimenter's work is very important here. Experiments could also be (and often indeed are) criticized for being unrealistic, as they do not fully reflect the natural setting. For example, a subordinate male fish might prefer large females over small ones in a simultaneous choice test, yet in nature this could be irrelevant if large fecund females tend to be paired to dominant males who defend their territory (and mate) vigorously. The defence of the experimenter to this challenge is similar to that of the modeller, too: the whole point of the experiment is to tell us that, *all other things being equal*, the large female is preferred over the small one. Likewise, the hare–lynx dynamics assumes that, all other things being equal, more lynx are bad news to hare populations, and this can run cyclic dynamics.

But there is an important distinction to be made between experimental approaches and modelling studies. An experimentalist is trying to find out if something (say, a male preference for large females) really does occur in nature. A modeller will never achieve this, which is good news to field ecologists were they scared of modellers taking over the scientific world. Models do not investigate nature. Instead, they investigate the validity of our own thinking, i.e. whether the logic behind an argument is correct. Are you interested in finding out if the simple fact that predators eat prey can *in principle* lead to population cycles? If so, go and model it. But whether this fact actually does cause cycles in real populations in Canada or Scandinavia or the vicinity of Kuopio must be investigated out there in nature – perhaps by removing predators from an area, or seeing if cycles are more often seen where the assumptions of this model appear to be better met, or by some other clever methods that the author of this book is not an expert on.

Once one begins to think of models as 'thinking aids' rather than investigations of natural phenomena, one could even go as far as to say that we only need models because our brains suffer from too many limitations and are not able to consider all sides of a complicated argument in a balanced way. Take, for example, R. A. Fisher's (1930) idea that sexually selected traits such as extravagantly long tails in birds exist because the following has happened. Initially a longer than average tail could have conferred a viability benefit to the male, perhaps because it improves flight performance. This means that females who prefer long-tailed males as mates will have offspring who have inherited a nice long tail. The genes for female preferences have now become statistically linked with longer tails in males, and long-tailed male offspring now have the additional advantage of being preferred by females. So, even if too long tails no longer give the viability benefit but instead diminish male offspring survival, the system may end up in a 'runaway' where both preferences and tail lengths evolve to ever more extreme values... Is it all now crystal clear in your head?

Don't worry, the confused feeling is shared by countless others who have read Fisher's account of the process. If only we all had brains capable of mathematically accurate split-second imagination, keeping precise track of all the relevant pros and cons of these genes – then Fisher would just have had to state his idea, and every person in the world would instantly have seen his point. Perhaps Fisher was like that, but most of us are certainly not. We rely on intuition and common sense, both of which can sometimes perform dismally badly.[1] This is why we

---

[1] For anyone interested in a wonderful account of mistakes of reason that are very hard to get rid of: read the book *Inevitable Illusions: How Mistakes of Reason Rule Our Minds* (Piattelli-Palmarini 1994).

end up arguing about whether from X really follows Y or not, and it is the same reason why the modelling niche in ecology is such a good one.[2] Modellers are not more intelligent than the rest, instead they simply gain their living by having learnt a mindset (plus a few mathematical tricks) that allows them to break the problem down into small pieces. This makes the assumptions explicit, and then one can derive the outcome in such small steps – often taken care of by a computer – that they, and everyone else, can trace the steps without suffering a brain overload like that suffered when encountering Fisher's work for the first time.

So, models only exist because we need them to help us: none of us are born with such supercomputer brains that we could evaluate arbitrarily complex arguments immediately and without external help. What is the optimal complexity of a model, then? Once again, it depends on the question. Reflect for a moment that there are maps with different scales. In the context of scientific models, it is useful to be reminded of the ultimate reason we do science: it is the joy of understanding something. If we could visualize and memorize much more detailed maps than we currently do, useful maps would include more detail than they currently do. Exactly analogously, if we could grasp much more complex processes without getting headaches than we currently do, models would look different too. Given the way our brains are built, a good guideline is that a model should include all the relevant details for the particular question at hand, but it should be kept so simple that it can be understood (if with joy, then still better). In other words, a model is not particularly helpful if it predicts that under conditions A the animal should do X, while under conditions B it should do X 30% of the time and Y in the rest of instances, and then there are 17 other parameters that interact with each other in producing a diversity of outcomes – but when asked why the model produces these effects of A and B, we still have no answer that can be expressed in a language that anyone's intuition can understand. Removing some additional detail from the model can then be surprisingly helpful: the effects of A and B could still be the same, but with far simpler equations.

For example, we might have spent a lot of time modelling the distribution of body condition in a population of migratory birds, ending up with very cumbersome equations, when a far simpler way to grasp

---

[2] Endless collaboration prospects – so you can pick the ones in the countries with the best food and weather. Field trips are to sites with already nicely established facilities, can usually be arranged in the best time of the year too, and someone else is doing all the tedious aspects of the data collection. Cool.

the conceptual issue is to divide up the population into two classes of individuals, 'hungry' and 'satisfied'. To show a conceptual point, this might be sufficient. Results could be far simpler to derive this way than with a more complete model, and if our understanding of the biology advances faster this way, the simplification is justified. But how to know, then, that the division has not caused some artifacts? Perhaps an exact shape of the body condition distribution would have produced a totally different answer? The answer is . . . we don't really know, unless we build the more complicated model too. (Which means that modellers rarely run out of models to study.[3]) Alternatively, it is often the case that the simple model has dealt with most of the thinking load, so that extrapolating to the last step can (fairly reliably) be achieved using imagination and verbal argumentation. This may sound unsatisfactory, but at least it provides a reason why any modelling quest should start with a fairly simple setting: effects of new added interactions are hard to judge otherwise.

In the above – and indeed in the rest of this book – I am mainly dealing with conceptual models. They are typically models that aim to answer questions, 'Does from A follow B?' Or in a little more complicated way: 'Under which biologically relevant conditions can we claim that from A follows B?' Such models usually aim at relatively broad taxonomic applicability, which also means that details of the behaviour of a specific species, no matter how exciting and important they appear, should usually be considered irrelevant. A modeller should not necessarily be judged as arrogant or ignorant if she brushes over such detail. When we want to know under what conditions, in general, one expects female preferences to evolve based on indirect genetic benefits, it is not very wise to consider details of the energetics of black grouse leks. Predictions of conceptual models, therefore, tend to be qualitative rather than quantitative. For example, we may predict that sexually selected male traits can be very costly to their bearer, whereas large costs of female preferences are not expected (because if we assume such costs, preferences evolve to zero, see Kokko *et al.* (2006a) for a review). The model is not intended to predict exactly how costly the trait will be in the case of a male black grouse.

---

[3] There is a real danger of becoming addicted to a problem. This leads to building towers of models on top of each other, adding this or that feature, until the meaning of it all becomes totally obscure to any outsider. If you intend to pursue the modelling path seriously, beware of this danger. It helps to go to conferences and listen mistakenly to talks of the session you did not intend to attend.

Instead, a good model should in the end say: these are the conditions that have to be met before we believe we have a logically consistent argument that explains male trait evolution; now go and find out if it really is the case. The need to test assumptions and not just predictions of a model can hardly be overemphasized. But as we have just learnt, assumptions never reflect the system completely faithfully, and that is perhaps the most important reason why an empiricist should have some grasp of mathematical modelling, even if not interested in pursuing it as a career. A model that is based on the assumption that the moon is a flat Roquefort cheese is obviously so out of this world that any conclusions drawn from it, no matter how mathematically solid, will have no relevance whatsoever. But a good and useful model will still appear 'false' in the sense that many aspects of reality are necessarily ignored, and here one must be able to judge if the assumptions nevertheless capture the essence of the biological argument.

Not all models refrain from making numerically explicit predictions, however. Some are much more applied and number oriented; perhaps we are not interested in the causalities at all, but we would really like to predict the number of rats left in an area after an eradication programme, extrapolating from past knowledge of rat behaviour. In such a case, it is probably not particularly important to know exactly why rats retain memories of bad food experiences for as long a time as they do. We can simply assume this happens, take an exact numerical value from experiments and build this into our model. Such models are predictive and may be precise (if we are lucky), but they do not try to aid our conceptual understanding of evolutionary processes. Of course, if we are really lucky, we might get at that too; some famous success stories where evolutionary insight combines with numerically accurate predictions come from studies that link sex ratios to local mate competition. There are some others too.

There are also models that fail to be simple enough for us to understand fully their inner workings when they churn out results, yet they are not precise either in the sense of predicting the numbers of rats on an island. Often such models are complex computer simulations, tracking the state and behaviour of a large number of virtual individuals, and examining the emergent properties of the system. Such results can be fascinating to watch – who wouldn't like to play god, or have their own study species repeating their fascinating behaviour on a computer screen – but I would like to warn against the overuse of such methods. It is quite easy to become lost in the virtual world: the computer will be happy producing results night after night, and the data can be summarized in a vast number of graphs – but did we really gain an understanding of why this or that curve went up or down when varying one of the umpteen parameters? Rarely. Very complicated individual-based models are perhaps most useful if they try to answer a question such as 'X can happen.' For example, when introducing spatial structure, altruism can spread in local subpopulations; or, an emergent property does exist that is interesting in its own right. They are useless when trying to show that 'X cannot happen', because this would require simulating an infinite number of possible parameter combinations. Not being causally transparent, the outcome of such models could, in the worst case, increase the thinking time required to understand the matter at hand.

Finally, there are models that, for lack of a better term, I call 'mathematically beautiful' models. These are often published in journals that empiricists rarely read, and they look particularly scary to the average ecologist with their theorems and proofs. This book will not cover them. It is good to realize, though, that many of them exist for a rather different purpose than advancing the understanding of a biological system: it is the advancement of mathematics per se. The authors of such papers typically share a love for mathematical beauty, which is a powerful concept that is exceptionally hard to define and not immediately clear to anyone who finds integrals petrifying. Nevertheless, biology abounds with processes that inspire mathematical thinking. Naturally, in an ideal world there should be free flow of information all the way from the extremely theoretical and beautiful results to the ugly and messy nature of ecological data. In practice, there are trade-offs: sometimes mathematical beauty has to be sacrificed for biological relevance, or vice versa. Also, some researchers find the views of the 'other side' fairly incomprehensible. Some gentle advice for the readers of this book with a very strong sense of mathematical beauty: some forgiveness in this respect usually helps a lot if your goal is to make biologists listen. And for the majority of behavioural ecologists: stop being afraid. Nobody masters every mathematical technique; therefore, do not ever be afraid of asking stupid-sounding questions, or beginning your own modelling career somewhere – a paper that maximizes the scariness of the squiggles involved is never a goal in its own right.

Finally, before turning to the actual modelling techniques, it is worth asking yourself: why do you want to build a model? If it is just to have a modelling paper on your CV, or that your system has not yet been modelled and models somehow seem to be valued by the scientific community – think again. Modelling without having a focused question in mind brings you nothing and blocks your computer from doing something more useful for you. But if you do have the question, which should sound something like 'I would really like to know if I am right in thinking that X can cause Y, because it tends to increase the availability of resource Z,...', then you are certainly on the right track. Please read on.

2

Population genetics

where we find males that treat females quite badly,
and some salmon get caught.

What is evolution? A standard definition is 'change in gene frequencies over generations'. Not all evolutionary change results from natural selection: a variety of processes can lead to changes in gene frequencies. For example, we could be interested in the effects of genetic drift – random changes in gene frequencies because of chance events that allow some individuals to reproduce, whereas others do not.

But more often than not, we are interested in the effects that *selection* has on gene frequencies. Charles Darwin's insight was that a trait that makes the organism survive or reproduce better will spread in a population – assuming it is heritable. Darwin knew nothing about genes, but using today's terminology, variation in the trait is caused by a gene having alternative forms called alleles. Genes are passed from parents to offspring, which results in inheritance of beneficial traits. Of course, a beneficial allele could be lost from a population through, for example, simple bad luck (more scientifically known as genetic drift). Even so, one can argue that if individuals who have allele A leave, *on average*, twice as many offspring as those who have allele a, then we would expect a tendency for the proportion of A to increase. Drift cannot always be ignored, but if we are interested in long-term evolutionary trends (explaining what can exist in nature) rather than a particular case of short-term change over a few generations, we could convince ourselves that nature will eventually reward alleles that have positive effects. We then allow ourselves to talk about the expected frequency of A increasing every generation, until it gets fixed in the population.

But this is not too exciting or motivating when it comes to learning how to model evolutionary processes. No-one needs a modeller to

13

become convinced that a trait that improves reproductive rates twofold should spread in a population (although one might have to resort to modelling when trying to get grips with the question of strength of drift versus selection, e.g. Falconer and Mackay (1996)). Let us, therefore, start our modelling tour with a relatively simple situation that is nevertheless substantially more complex than 'twice as fast reproduction', and hope for some biologically interesting outcomes.

Figure 2.1 shows a weird biological structure: the penis of a seed beetle (bean weevil) *Callosobruchus maculatus*. One cannot help noticing that it is rather spiky, and indeed, male weevils puncture female tissue during copulation (Crudgington and Siva-Jothy 2000). Females do not seem to like this, but kick vigorously during copulation. The damage is not insignificant: in an experimental setting, doubly mated females died significantly younger than singly mated females (Crudgington and Siva-Jothy 2000).

Why should a male be so mean to the females he is mating with? Crudgington and Siva-Jothy offer two possible explanations. If mating is dangerous to a female, she might become less willing to remate, which is obviously good news to the male who is interested in protecting his paternity. Or, if damaged females have lower survival, females might respond to genital damage by increasing their oviposition rate, following the logic 'if you have little time left to survive, better invest a lot in current reproduction'. Either way, damage to the female can mean more offspring sired by the male.

This phenomenon is an example of *sexual conflict*. The battle can take many forms, from sexual harassment in wild sheep, which can lead to female death (Reale *et al.* 1996), to male snakes sometimes causing female death through suffocation (Shine *et al.* 2001), and to toxic chemicals carried in seminal fluids in *Drosophila* (reviewed in Chapman *et al.* (2003) and Arnqvist and Rowe (2005)).

But should we *really* expect evolution ever to favour traits that diminish the reproductive output of females? Females are the sex that actually produces the offspring of the next generation. Imagine meeting a student who argued that any gene that harms female reproduction should be very strongly selected against, even if it is advantageous to males. After all, males are often replaceable: if one male does not fertilize a batch of eggs, there will be another one available.

How would one respond to such a claim? One argument is that evolution can very well favour 'selfish' genes (Dawkins 1976), so even if the male trait hampered the production of future generations, one could

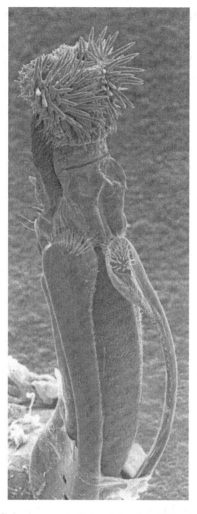

Fig. 2.1 Penis of the bean weevil (seed beetle) *Callosobruchus maculatus*. (From Crudgington and Siva-Jothy, 2000; reproduced with permission from Microscopix Photolibrary (http://www.microscopix.co.uk); © Andrew Syred.)

imagine it taking over. Yet, the fact that the male could be shooting himself in the foot does sound like it should be taken seriously: after all, he is diminishing the reproductive output of his own mate. Maybe you are beginning to doubt which side should win? Whenever we have a case where our brainpower lacks the bandwidth to see through the argument so clearly so that there is no disagreement left, it is time to resort to some modelling.

2.1 A population genetic model of sexual conflict

What do we need to build our first model? Firstly, be clear about what the question is. In this case: can a heritable trait spread if it aids male reproduction at the expense of female fitness? Good. Defining the problem was easy in this case, but what next?

Remember this rule: never make things more complicated than necessary to prove your point. Given that we want to follow the spread of a trait, it is probably simplest to begin with just two possible forms (alleles) of a gene. This leads us naturally to the realm of population genetics, which is the study of gene frequency changes, usually considering a small number of different alleles that are the targets of selection. In population genetics, two alternative alleles of a trait are often denoted A and a, so let us adopt this notation too.

Here, let us get sidetracked for a moment. Notation is all about communicating ideas. If you wanted to call the alleles ζ and ψ, or Bob and Fred,[1] you could do so without changing the truth value of your model at all. But models often end up having many parameters and variables.[2] Simple and logical notation could make all the difference between achieving that 'eureka, I understand!' moment or getting hopelessly stuck. The same applies to others who you are trying to convince that your argument is correct. This is why it is sometimes worth stopping and thinking for a while before choosing labels for traits and events. In our case, A and a are quite nice choices because the A allele makes males more 'Aggressive' when competing for females.

There are some conventions that people use and that one will learn when reading enough modelling papers – but one also quickly notices that there is no universally adopted dictionary for names of variables. For example, models that consider something that changes over time usually use t to denoted time, but in a paper on sexual selection t could

---

[1] Multi-lettered variables are best avoided when writing up your results because '*mad*' could mean a single variable describing aggressive behaviour, but it could also be understood as a multiplication, m times a times d. That said, multi-letter names within a computer program can make the code easier to understand. In computer languages, there is no ambiguity involved, because multiplications are made explicit with a symbol, usually * (some programs also accept a space). Therefore, it is quite common and entirely normal to end up with slightly different notation in the program code and the glistening end product, which is the scientific publication.

[2] *Parameters* (sometimes called *control parameters*) are things that you give fixed values (for each scenario you study), e.g. m and b in the current example will be parameters. What will be denoted by y, by comparison, is not a parameter but a *variable*; one does not pre-set it to any particular value, but its value and evolution over time is computed based on the values of parameters.

denote the mean value of the male trait (e.g. Iwasa *et al.* 1991). So, if reading a paper, try not to freak out even if you encounter a bunch of ζs and ψs: squiggly as they might look, they are never anything more complicated than labels for, well, numbers. Modelling never means more than taking complicated arguments and reducing them to the simplest building blocks, and then seeing what happens when these interact. Trying to adhere to rules that allow everyone to communicate to each other more fluently is recommendable – but at the same time, do remember that there are no rigid rules except that notation should be consistent within the same study. This is why you should feel free to choose unusual notation if it really does help the reader understand what you are talking about. But be consistent: letting t indicate time on page 6 and mean trait value on page 7 of a manuscript or study report is certainly a very bad idea. In this book, I have tried to stick to consistent notation within each chapter. Between chapters, however, notation will vary. This is partly because different methods tend to use the same letter to denote different things, but partly because it is a fact of (a theoretical ecologist's) life that there is no such thing as a comprehensive list of symbols for biological phenomena that everyone adheres to. Life skills of anyone interested in reading or writing theoretical papers must, therefore, include the ability to speak about x and y in a wide variety of meanings.

Back to our aggressive (A) or not so terribly aggressive (a) alleles. How can we know if A will spread? Somehow we have to be able to predict its frequency from one generation to another. Because generations follow each other in time, let us denote generations with t; so the initial starting point of the population occurs when $t = 0$, after one generation we have $t = 1$, and so on. (Nothing prevents you from specifying that the start happens at time $t = 1$; choose whatever feels most logical to you.) The variable that we are really interested in is the frequency of A at each point in time, $x(t)$. This notation means that x is a function of time t, in other words it takes different values for each value of t. For example, we could have $x(17) = 0.3$, and $x(18) = 0.34$, if the frequency was increasing from 30% to 34% between generations 17 and 18. Why do we use x to denote allelic frequencies? No particular reason. Most people remember from school (or from movies) that the letter x carries the vibe of being 'unknown', and we do not know yet how the frequencies will evolve – hence x.

What next? We now want to track evolutionary change. This grand statement means nothing more than writing down an expression for $x(1)$, assuming that we started from $x(0)$. Or, more generally, we are asking what is the frequency in the next generation, $x(t+1)$, if the current frequency $x(t)$ is known.

In order to get the number $x(t+1)$, we need to do some maths. This is no scarier than doing some bookkeeping of offspring numbers. To make any equations look less scary still, let us simply drop the indexing with time (remember, you are allowed to do anything with notation as long as it does not become ambiguous), and write x when we really mean $x(t)$, and x' when we mean the frequency in the next generation, $x(t+1)$. Therefore, the problem is the following: if the initial frequency of the allele is x, how many matings will there be between males and females having either allele A or allele a? How many offspring will be produced in each of these matings? What alleles will the offspring be carrying?

We shall now tackle each of these questions in turn. Regardless if female or male, an adult individual will have the A allele with probability x, and the allele a with probability $1-x$. What sort of matings are possible in the population? It could be that the mother is a, and the father is a; or the mother is a, and the father is A; or mother is A, and father is a; or both could be A. Imagine walking around the population during the mating season and making notes of the relative frequencies of each type of mating. These are probably going to be important for seeing what happens next, so we give them labels – for example, $p_{a \times a}$, $p_{A \times a}$, $p_{a \times A}$ and $p_{A \times A}$. The letter p here stands for 'probability'.

Hang on! In the above, we were assuming haploid genetics, which means that females and males have either allele a or A, but never both. This does not take into account that there are two copies of the same gene in any diploid organism. It is a deliberate choice here, to keep the number of different types of mating down. We can probably justify this, by arguing that diploidy is not really essential for the question we are studying. If A is advantageous enough in sons to counteract the harmful effect it has on his mate, then we would also expect AA individuals of a diploid species to perform better than aa, while the heterozygous Aa types would simply have some kind of effect that resemble either AA, aa, or something in between. For example, it could be that allele a was dominant so that Aa individuals behave like aa. Some A alleles therefore become 'invisible' to selection in a diploid species, and this could slow down the evolutionary process. Nevertheless, if the net effect of A is beneficial so that A (haploid) or AA (diploid) individuals on average have higher fitness, the trait will eventually spread. This is why we can argue that our haploidy assumption is not really making the analysis invalid even for diploid species. In any case, however, you need to be aware of the assumptions inherent in a model; sometimes researchers do forget to be explicit about them. (Think for a moment about this problem. What other hidden assumptions have we made? Do any of them worry you? Find the answers at the end of this chapter.)

Still, one must remember that the real reason for us to ignore diploidy is nothing more glamorous than our reluctance to deal with too many complications: haploidy makes the bookkeeping so much less tedious. But maybe the above verbal argument did not convince you that diploidy could not make a huge difference. What to do then? The basic remedy for not knowing if a logical argument holds or not is, once again ... to model it. So, if you think diploidy could cause a deviation from what happens with selection under haploidy, there is no other way to proceed than to look at the difference and repeat all the modelling for a diploid organism as well. This requires more work than the modelling done here: firstly because there will be many more types of mating pairs (e.g. $AA \times Aa$), and secondly, one should probably look at a variety of different possible patterns of allelic dominance. For an example of how such work can nevertheless lead to graphs that are equally simple as those presented here, see Gavrilets and Rice (2006).

But let us decide that selection in a haploid organism is not conceptually different from one in a diploid one for our current purpose.[3] In particular, we are interested in whether it is at all possible that evolution can favour traits that harm female reproduction simply because they give some males an advantage over others. If we can show that this can happen in our haploid organism, we know that it is possible in principle. We now get back to the four proportions $p_{a \times a}$, $p_{A \times a}$, $p_{a \times A}$, and $p_{A \times A}$, which describe the relative frequency of matings between each type. How do they depend on x? For a moment, let us ignore the mating advantage of type A males, and consider what happens if matings occur completely randomly.

If a fraction x of both males and females have allele A, then the relative frequency of matings between an A female and A male is $p_{A \times A} = x^2$. How come? It follows from the basic ways in which probabilities behave. If Joanna has a blue bag with 3 oranges and 7 apples, and Judith has a red bag that also contains 3 oranges and 7 apples, and each of the girls chooses one fruit blindly from each bag, then the probability that both picked an

---

[3] One could have chosen differently: it is even possible to build models that track the evolution of haploidy versus diploidy or that of dominance, see Chapters 8 and 12 in Otto and Day (2006). It is up to the modeller to decide what is fixed, and what is free to evolve. This decision depends on what the question is, which may sound utterly trivial. Yet, being able to articulate the appropriate question is a large part of skilful modelling.

orange is $3/10 \times 3/10 = 0.3^2 = 0.09$. Likewise, each mating in a randomly mating population means looking at a randomly chosen female – who is type A with probability x – and choosing a male for her, again being type A with probability x; hence x^2.

But the whole point of our model is that mating success of males is not random with respect to the genes they carry. Instead, we assume that A males are more competitive over matings. Now, the details might matter: if their advantage relies on a faster oviposition rate (which makes it less likely that a female has the time to remate before all 'his' eggs are laid), the exact pattern of paternity could differ somewhat from a case where the advantage is a refusal by 'his' female to remate. However, our aim is not to model the specifics of the *C. maculatus* system, but to ask a very general question about whether male-induced harm can evolve in principle. It is, therefore, best to aim for simplicity: let's say that for every batch of eggs sired by an a male, the A male sires m times as many eggs, and let's also equate a 'batch of eggs' with the outcome of one mating. However complicated the patterns of sperm storage and competition might be, we choose to step over such detail. The letter m denotes a parameter (see footnote 2 (p. 16)) that specifies the strength of the mating advantage conferred by allele A. If $m = 1$, A does not bring any advantage over a males, while values of m much greater than 1 mean that A has a very strong advantage.

If $m > 1$, then $p_{A \times A}$ no longer equals x^2. Females all get mated regardless of their genotype, so the probability that a mating female is A still equals x. But when $m > 1$, the probability of a mating male being A exceeds their proportion x. Let us say y equals the probability that a *mating* male has the A allele, and $p_{A \times A}$ then equals xy. Males with allele A are disproportionately more 'prevalent' in the pool of mating males, which leads to

$$y = \frac{mx}{mx + (1 - x)} \tag{2.1}$$

Where did this equation come from? To resort back to the example of apples and oranges: if oranges are m times as sticky as apples and therefore more likely to be picked, their 'effective number' in the bag becomes $3m$, while the number of apples stays at 7. Therefore, the effective number of oranges to be chosen is $3m$, and the total effective number of fruit is $3m + 7$. Thus, the proportion of times that an orange is chosen is $3m/(3m + 7)$.

It is instructive to see what y looks like as a function of x, with different values of m (Fig. 2.2). If you have access to any computer program that handles mathematical expressions, you can make use of it: the examples

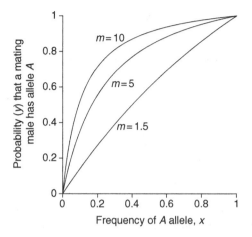

Fig. 2.2 The relationship between the frequency of the *A* allele, *x*, and the probability *y* that a mating male has this allele, for three different values of *m*.

of this book use MATLAB notation (Box 2.1), but all that is essential is the line y=m*x./(m*x+1−x). This tells the program to calculate *y* for each value of *x*, according to Eq. (2.1). The rest of Box 2.1 contains instructions to consider particular values of *x* and *m* and commands that order plots on the screen; these typically vary from program to program but are rarely difficult to learn. Box 2.1 also includes instructions for a spreadsheet version of the calculation, and a pocket calculator would do the same, just a little bit more tediously – one could then plot the relationship on a piece of paper, point by point, using a pencil and a ruler. That's all the magic there is.

Box 2.1

How to create the curves of Fig. 2.2 in MATLAB. The rows that begin with a % symbol are explanatory only: MATLAB does not need them. The following rows can be straightforwardly typed in the command window of MATLAB. Alternatively, the lines can be saved in a file called, e.g., showy.m (for 'show y'), and afterwards the plot of *y* against *x* can be created by typing showy in the command window. MATLAB's current directory must match the location of showy.m for this to work.

```
% 'linspace' can be used to make a vector with
% linearly equally spaced values
```

Box 2.1 cont.

```
x=linspace(0,1,101);
% now we choose a value of m to be investigated
m=5;
% calculate how y depends on x
y=m*x./(m*x+1-x);
plot(x,y)
xlabel('Proportion of A males in the population')
ylabel('Proportion of A males among mating males')
```

What do these lines do? MATLAB records everything as a number, and these numbers may be arranged in rows or columns called *vectors*, or arrays of values called *matrices*. Here we first create a vector x that takes 101 values between 0 and 1, in other words, 0, 0.01, 0.02, ... , 1. (After x has been defined, one can ask MATLAB to show all these values by typing x into the command window and pressing enter.). Then we choose a value of *m* to be investigated, after which we are ready to calculate *y* for every one of the 101 values of *x*, according to the relationship specified in Eq. (2.1).

In a less refined way, one could create a column of 101 values for *x* in a program such as Excel, and calculate *y* in a different column. The choice of a program to use does not matter in principle, but some programming languages are more tedious than others to use when building more advanced models. To some extent this is a matter of personal preference (and, quite often, a considerable amount of developmental inertia).

To plot all three lines in Fig. 2.2, we will need three sets of values for *y*. It is up to you how to name them, and one way that works is this:

```
x=linspace(0,1,101);
m1=1.5; m2=5; m3=10;
y1=m1*x./(m1*x+1-x); y2=m2*x./(m2*x+1-x);
y3=m3*x./(m3*x+1-x);
plot(x,y1,x,y2,x,y3)
xlabel('Proportion of A males in the population')
ylabel('Proportion of A males among mating males')
```

The probability that a randomly chosen mating male is of type *A* (which is, of course, equal to the proportion of *A* males among mating males) is larger than their proportion in the population when $m > 1$, and the difference increases with *m* (Fig. 2.2). However, note that the effect of *m* depends on x. For example, when $m = 5$, the frequency of A males in the mating population is generally *not* five times their frequency in the whole population. This would clearly be wrong. If, say, 80% of males are A, it is impossible to have $5 \times 0.8 = 400\%$ of males being A when mating. The curves in Fig. 2.2 'know' this and automatically adjust for the proportion of A males present: everyone being a means $x = y = 0$, and if everyone is A, both x and y must of course equal 1. Everything works out nicely because Eq. (2.1) gives each A male a fivefold advantage over a males, not over *average* males.

In general, it is useful to check assumptions in a model by plotting relationships between variables with various parameter values. If we had been a little too quick and assumed $y = mx$ instead of Eq. (2.1), looking at a figure with y shooting through the 100% roof (try this out!) would have set the alarm bells ringing.

Now that we know the proportion y, we can work out the proportions of each type of mating:

- both female and male are A with probability $p_{A \times A} = xy$
- female is A, male is a with probability $p_{A \times a} = x(1 - y)$
- female is a, male is A with probability $p_{a \times A} = (1 - x)y$
- both are a with probability $p_{a \times a} = (1 - x)(1 - y)$.

We can now plug in Eq. (2.1) to replace y in these expressions. This allows us to know how probabilities are derived from just knowing x and m. For example, $p_{A \times A}$ becomes $x[mx/(mx + (1 - x))]$. If you prefer, this is the same as $mx^2/[1 + (m - 1)x]$. Algebraic simplification of expressions can be useful because the outcome often looks much more tractable, although the difference is barely perceptible in this case. Now is a good opportunity to practise these skills: write down the expressions for $p_{A \times A}$, $p_{A \times a}$, $p_{a \times A}$ and $p_{a \times a}$, and calculate their sum. If you have not made a mistake, the sum must equal 1. Why? Every mating pair must be one of the four mutually exclusive types, there simply is no other possibility. Therefore, if the probabilities do not sum up to 1, something is badly wrong.

Sadly, we are not finished yet. What do we need to know to get from matings to offspring production? Clearly, one thing we need are the genotypes of offspring that result from each type of mating. But we should not forget one of the assumptions we wanted to make: that females who mate with somebody having the allele A should have reduced fitness. Now, what is fitness?

Fitness is the expected genetic contribution of an individual to future generations. Female fitness could be harmed, for example, by shortening her lifespan, making her less able to care for her offspring, reducing the number of eggs she can produce, and so on. Now we realize with mild horror that we have not thought at all about the details of the life history of our species. Should we imagine that female lifespan decreases if her mate has allele A? But we have not specified lifespan in the first place. Our intention was probably to count the number of alleles in the offspring and that's it when it comes to changing from one generation to the next. So it is probably simplest to think about *non-overlapping generations* in which everyone dies after reproducing. Otherwise we would have to look at survival from one day or year to the next, think about population structure with different-aged individuals who have derived their genes from a different gene pool, and so on. There is no reason why these details should be central to our question: can the male advantage win over the female disadvantage?

Non-overlapping generations it is, then. (Remember to write this down as one of the assumptions you are making.) This means that it now makes sense to concentrate on a case where females suffer reduced fecundity as a result of mating with harmful males. It could of course be that female lifespan is reduced and therefore she does not have the time to lay all her eggs, but the effect is the same: 'fewer eggs laid' and 'lower fecundity' are, effectively, the same thing. And how much do we want her to suffer? Let's be general and consider all possibilities. Let her fecundity be a fraction b of what she would have produced otherwise, and think for a moment about sensible values of b; clearly the value of b must be between 0 and 1 (to be exact, $0 \leqslant b \leqslant 1$; as a borderline case, when $b = 1$, the female's fecundity is not harmed at all). And what does she produce otherwise? That can be called B: it can be thought to equal the 'Batch of eggs' that we decided to assume is the outcome of each mating. Mind the language, however: there is no particular assumption here that eggs are laid in batches, we're simply counting the offspring that result from a mating.

The fraction b might appear a complicated way to express a cost. Why not say that the female suffers a cost C and lays $B - C$ eggs where B is her potential maximum fecundity? Two reasons. The magnitude of a value of C, say $C = 3$, would be hard to interpret. It might be a very large cost if $B = 5$, or a minute cost if $B = 3000$. Secondly, one could easily overlook logic if running through a large set of values of B and C, perhaps ending up with silly combinations where $C > B$, and the fecundity of females mated to harmful males is negative.

Are you ready? How many A offspring will there be in the next generation? First give this quantity a name, for example n_A. Some of these offspring originate from $A \times A$ matings (all offspring from such matings will be *A*), but some from *A*×*a* or *a*×*A* matings (half of offspring from such matings will be *A*, remember the rules of haploid genetics). We also need to keep in mind the number of different types of mating couple, and the fact that fecundity is reduced to a factor *b* of the original if the *male* is A. Female genotype does not matter here as it is not expressed. So if the total number of all mating pairs is denoted by *N*, we get

$$n_A = N\left[p_{A\times A}bB + \frac{1}{2}p_{A\times a}B + \frac{1}{2}p_{a\times A}bB\right] \tag{2.2a}$$

If this was too quick, have a look at it again. The total number of matings *N* is multiplied by the expected fecundities from each type of mating, weighted by the frequencies of each type of mating pair (in modelling terminology, 'weights' capture what was meant by the apples and oranges in the fruit-picking example). Alternatively, *N* can be expressed inside the bracket, which implies that there are $Np_{A\times A}$ matings in total of the type *A*×*A*, each of them giving a fraction *b* of the *B* offspring that the female could have produced had the male been *a*; and there are $Np_{A\times a}$ matings of the type *A*×*a*, each now yielding *B* offspring but only half of them carry the allele *A* . . . , and so on. Deriving the number of offspring carrying the *a* allele follows similar arguments and yields

$$n_a = N\left[\frac{1}{2}p_{A\times a}B + \frac{1}{2}p_{a\times A}bB + p_{a\times a}B\right] \tag{2.2b}$$

Note that the fecundity reduction in offspring production applies regardless of the allele that the offspring inherits; it is a phenomenon that depends on the allele expressed by the father.

Where are we now? We know how to get from *m* and *x* to $p_{A\times A}$ and the other probabilities. We also know how to get the number of offspring of each allele, n_A and n_a, once the probabilities are known. The frequency of *x* in the next generation, which we chose to call *x′*, is now almost within sight. If we were producing, for example, 2000 offspring with the *A* allele and 1000 with the *a* allele, the correct proportion of *A* is 2000/(2000 + 1000) = 2/3. (In general, always try out some simple numerical examples if you feel stuck.) The idea is the same for all values of n_A and n_a, thus we have $x′ = n_A/(n_A + n_a)$.

So we have our model! This is now something we can play with: from *x* we get to *x′* so we can test if the allele *A* spreads or not. We can do this as long as we know where we start (*x*), and the values of *N*, *b*, *B* and *m*: *x* and *m* will give us *y*; then *x* and *y* will give us the mating probabilities. We then use Eqs. (2.2a) and (2.2b) to get n_A and n_a, and the rest is a simple

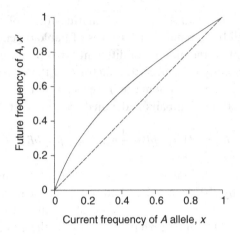

Fig. 2.3 The solid line indicates the relationship between the current and future frequencies of the allele A, when $m = 5$ and $b = 0.9$. The future frequency is higher than the frequency indicated by the diagonal (dashed line). Since the diagonal marks cases where $x' = x$, the difference indicates that the frequency of A is increasing throughout the range of its current frequencies, except where $x = 0$ or $x = 1$.

division. But how do we know the starting values for N, b and so on? The answer is that we are free to explore any combination. They are parameters of the model so we can play with them freely as long as they make biological sense: values of b that we explore should lie between 0 and 1, for instance.

What do we find when calculating the shift from x to x' (see Box 2.2 for how to do this in MATLAB)? If m is really quite large (e.g. 5; remember that $m > 1$) and so is b (e.g. 0.9; remember $0 \leqslant b \leqslant 1$), we find a figure where the curve x' always lies above the diagonal, except if $x = 0$ or $x = 1$ (Fig. 2.3). This means that for each value of an allele frequency, the new frequency will be higher than the previous one. If $x = 0$, there are no A parents, and hence no A offspring can exist either. This is an equilibrium (meaning that this gene frequency will not change over time), but it is an *unstable* one: it can only be maintained if there are absolutely no A individuals present. As soon as x slightly exceeds 0 – which happens if A arises through mutation or through immigration – the new frequency will be higher in the next generation, higher still after that, and this does not stop until $x = 1$. In other words, A will spread to fixation, and $x = 1$ is the *stable equilibrium*, towards which evolution proceeds and returns even if perturbed away from it temporarily.

Box 2.2

Creating plots in Figs. 2.4 and 2.5 in MATLAB can be done with the following lines.

```
% make a vector of 101 different values of x, ranging
% from 0 to 1
x=linspace(0,1,101);

% choose a value of m to be investigated
m=5;

% calculate how y depends on x
y=m*x./(m*x+1-x);

% calculate mating proportions
pAA=x.*y;
pAa=x.*(1-y);
paA=(1-x).*y;
paa=(1-x).*(1-y);

% let's check that these sum up to 1!
pAA+pAa+paA+paa

% now to nA and na. We must also choose values for N, B,
% and b
N=1000; B=20; b=0.8;
nA=N*(pAA.*b.*B+0.5*pAa.*B+0.5*paA.*b.*B);
na=N*(0.5*pAa.*B+0.5*paA.*b.*B+paa.*B);

% then we get xnew, and can plot it against x
xnew=nA./(nA+na);

% then plot things: first xnew against x, and then x
% against x
% - the latter will produce a diagonal
plot(x,xnew,x,x)
xlabel('Frequency of A in generation 0')
ylabel('Frequency of A males among mating males')
```

Note that we did not here use the full equation of x' (in MATLAB, xnew) as defined in Eq. (2.5), although we could have done so. Instead, we first calculate the 'building blocks' such as $p_{A\times A}$, $p_{A\times a}$, and so on, which is probably the clearer way. This is because we can now have a look at the way $p_{A\times A}$ (or any other intermediate step in the calculations) depends on x, which we could not have done if we had used Eq. (2.5) for x'.

Box 2.2 cont.

The above is a *script* in MATLAB jargon. It is a quick and dirty way to see what is going on, but for results that are intended to be more permanent, it is generally better to write *functions* (see Lesson 2 in the Appendix for basic instructions). A function performs exactly the same operations, but it has clearly defined inputs (or *arguments*). This is a list of all the factors, typically the parameters of a model, that we want to vary to see their effects on the output. In our case, the inputs should be m, b, N and B, and perhaps also a technical detail which is the accuracy with which one wishes to compute the solutions. Above we used 101 different values of x, but perhaps sometimes we'd like to compute less (with a slow computer at a field site?) or sometimes more (e.g. when producing the final figures for a publication). The values for variables such as y and nA always follow using the same rules, so they don't have to be given as inputs and are instead taken care of by the inner workings of the function. Another advantage of a function is that intermediate and temporary results do not clutter the workspace of MATLAB but are deliberately 'forgotten' as soon as the function has been run.

```
function [xnew,x] = sexconflict(m,b,N,B,accuracy)
% function [xnew,x] = sexconflict(m,b,N,B,accuracy)
% Allele frequency change in one generation.
% Give m, b, N, and B according to the definitions in
% Chapter 2 and accuracy as an integer number that
% tells how many values between 0 and 1 the variable x
% should take.
% The function computes xnew and x.

x = linspace(0,1,accuracy);

% first calculate how y depends on x.
% Note there is no need to define a value for m now - it
% will have been supplied as an argument when using
% the function 'sexconflict'.
y = m*x./(m*x+1-x);

% now calculate mating proportions
pAA = x.*y;
pAa = x.*(1-y);
paA = (1-x).*y;
paa = (1-x).*(1-y);
% now calculate nA and na
```

```
nA=N*(pAA.*b.*B+0.5*pAa.*B+0.5*paA.*b.*B);
na=N*(0.5*pAa.*B+0.5*paA.*b.*B+paa.*B);

% then we get xnew and can plot it against x
xnew=nA./(nA+na);

% then create the plots: first xnew against x, and
% then x against x
% - the latter will produce a diagonal figure(1);
plot(x,xnew,x,x)
xlabel('Frequency of A in generation 0')
ylabel('Frequency of A males among mating males')
```

How to use this function? Any function will first have to be saved in a file before it can be used; here it can be called `sexconflict.m`. The file should be in the current directory where you are working in MATLAB. (It usually makes sense to create a directory for each modelling project you are working on.) Then you can plot a figure by typing the following line in the command window:

```
sexconflict(5,0.9,200,20,101)
```

where the first number gives the value of m we are interested in, the second one the value of b, and so on in the order defined by the first line of `sexconflict.m`. Try out lots of different values! If you like to store the results in various outcome variables for further comparison, that is easy too: here, we save results separately for Figs. 2.3 and 2.4.

```
% note that 'x' needs to be defined once only
[xnew_fig23,x] = sexconflict(5,0.9,200,20,101)
[xnew_fig24,x] = sexconflict(5,0.1,200,20,101)
```

Finally, try out

```
help sexconflict
```

to see that the first lines of the function, which begin with a % symbol, have become a neat little help feature: these lines are printed when asking for help with this function. This is also the reason the first line that contains the function statement appears twice: the first one gives the syntax for MATLAB, the second one for us, when we ask for help.

If m and/or b are lower, we end up with the opposite case (Fig. 2.4: here, $m = 5$ and $b = 0.1$). In other words, a fivefold mating advantage in males is not sufficient to compensate for a drastic 90% drop in female fecundity. The frequency of the A allele will decline, whatever its initial frequency, and the decline only stops at $x = 0$. Now $x = 0$ is the stable equilibrium, and $x = 1$ is the unstable one; allele A does not get fixed,

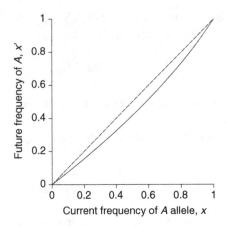

Fig. 2.4 The relationship between the current and future frequencies of the allele A, as in Fig. 2.3, but with $b = 0.1$. Now the future frequency is below the current frequency, indicating that the frequency of A is declining.

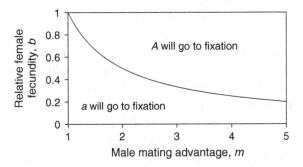

Fig. 2.5 Parameter regions that fulfil $(m > 1/b)$ or do not fulfil $(m < 1/b)$ the conditions under which allele A spreads to fixation.

instead $x = 0$ is approached from any frequency $x < 1$, no matter how close the initial frequency might have been to the value 1.

We can also easily explore the effects of other parameters. For example, we might be interested in the effect of population size (measured as the number of mating pairs, N). In stark contrast to the effects of b and m, we find that the figures do not change at all from Figs. 2.3 or 2.4, no matter what extreme values of N we try (verifying this can be done using Box 2.3). And, similar things happen if you try different values of B: no effect on the solution at all.

Box 2.3

A quick MATLAB check to see that population size N does not seem to matter:

```
N_case1 = 1;
xnew_case1 = sexconflict(5,0.5, N_case1,20,101);
N_case2=100000;
xnew_case2=sexconflict(5,0.5, N_case2,20,101);
```

To check that the figures really are identical, we can subtract the solutions of the two different cases from each other and look at the maximum absolute value of the difference between them. (The absolute value of 2 is 2, that of -2 is 2 too, so using absolute values avoids the problem that MATLAB might not think of large negative differences as being large.) This difference should be 0, to the limits of numerical accuracy, if there is no difference.

```
max(abs(xnew_case1-xnew_case2))
```

We have learned some things by now.

- It is possible, but not inevitable, that fecundity-reducing traits spread in the population.
- The spread has something to do with the values of m and b, but not with N and B.
- It looks like there is always fixation of either the a or the A allele, because the only points where $x = x'$ seem to be at $x = 0$ or $x = 1$.

The problem? We now have a fairly good feel of the system, but all our results are *numerically* derived. This is jargon, and it means that we have to generate each solution separately by running the program one more time. Wouldn't it be nice if we could be more certain: what if we could, for example, *prove* that only $x = 0$ or $x = 1$ are possible equilibria, and no intermediate gene frequencies exist where we expect no further evolutionary change? Or, what about finding out a general statement about how exactly m and b together determine which equilibrium is approached? The downside of numerical solutions is that while they can prove that something can happen (e.g. that m influences which equilibrium is reached), they cannot prove that something can't. For example, no matter how many different values of m, b, N and B we try out, we could always have overlooked some weird combination of values that could produce a surprise equilibrium where $x = 0.34$ is stable, or a case where N turns out to influence the location of the equilibrium after all.

This is the reason why *analytical* solutions are preferred over numerical ones. An analytical solution is a statement such as this: X happens if, and only if, $a/b > d \exp(-g)$. While this may look decidedly unfriendly, it is a very compact way of making general statements. The above (totally hypothetical) example is a mathematician's way of saying that increasing the value of a makes X more likely to happen as long as b is non-negative, and that if a and b are positive and d is negative then X is true regardless of the value of g. If you have reached an analytical solution, you can then interpret the findings biologically, satisfied with the knowledge that what you have just stated is absolutely, fundamentally, always true. No-one can ever turn up with a new value of d that destroys your whole argument. But before you celebrate too hard, remember that someone could still turn up with a complaint regarding the biological realism of the assumptions that have led you to your pathbreaking equation!

A Finnish proverb: salmon is so good it's worth fishing even if one fails to catch any. In the world of modelling, we cannot always be sure that we will find an analytical solution, and it may be tempting to be satisfied with figures such as 2.3 and 2.4: certainly we have already gained some understanding. But it's worth trying to catch the analytical solution – so let's try.

We know that $x' = n_A/(n_A + n_a)$, and the goal is to work until we have an expression that links x' directly to x and the parameters N, b, B and m. For example, we know that n_A depends on m because the probability that a mating male has the A allele depends on m, but this needs to be made explicit now. We have

$$
\begin{aligned}
n_A &= N\left[p_{A \times A} bB + \frac{1}{2} p_{A \times a} B + \frac{1}{2} p_{a \times A} bB \right] \\
&= NB\left[p_{A \times A} b + \frac{1}{2} p_{A \times a} + \frac{1}{2} p_{a \times A} b \right] \\
&= NB\left[xyb + \frac{1}{2} x(1 - y) + \frac{1}{2}(1 - x)yb \right]
\end{aligned}
\tag{2.3}
$$

Similarly for n_a, which results in (after a similar number of steps)

$$
n_a = NB\left[\frac{1}{2} x(1 - y) + \frac{1}{2}(1 - x)yb + (1 - x)(1 - y) \right]
\tag{2.4}
$$

The $x' = n_A/(n_A + n_a)$ now becomes something that looks rather nasty, and x' now equals (Eq. 2.5):

$$x' = \frac{NB[xyb + x(1-y)/2 + (1-x)yb/2]}{NB[xyb + x(1-y)/2 + (1-x)yb/2] + NB[x(1-y)/2 + (1-x)yb/2 + (1-x)(1-y)]} \quad (2.5)$$

and certainly this is not something that one would call simple and elegant. If 'horror' is a word that springs to mind, consider that all there is in x' is a summary of all the little steps of logic we have meticulously followed so far. If you can follow every step of book-keeping, you can write down Eq. (2.5) too. Published models unfortunately tend to keep some steps in between hidden to save space, but the principle remains the same: with patience one can work through all the details.

The other good news is that Eq. (2.5) can be simplified. Algebra is full of rules that tell what can be done to expressions so that their values do not change. For example, $a(b + c) = ab + ac$ is always true, and that's why the denominator, which looks like $NB[$squiggles$] + NB[$some other squiggles$]$, can also be expressed as $NB[$squiggles $+$ some other squiggles$]$. After this, one notices that both the numerator and the denominator are proportional to NB, in other words, both N and B cancel out. (Problems only arise if one of them is zero, but then we won't have a population at all.)

Neither N nor B can therefore influence x' at all. In fact, in this case we can catch quite a large salmon. After having got rid of N and B, more algebra is needed: surely some terms in Eq. (2.5) can be combined or cancel out. Try it – it will work, and eventually x' simplifies to

$$x' = \frac{1}{2}\frac{x + y[b - (1-b)x]}{1 - (1-b)y} \quad (2.6a)$$

Still, we would rather have m appear in the equation, rather than y. Equation (2.1) tells us what to substitute in place of y. The result initially looks like it has gained some nastiness again, but do not worry: this is temporary. By now we are getting familiar with algebraic simplification procedures, and we will find,

$$x' = \frac{1}{2}\frac{x(1 - x + bm(1 + x))}{1 - (1 - bm)x} \quad (2.6b)$$

From this we can proudly announce that we know that N and B will not change the course of evolution, in our deterministic model at least. The benefit (or cost) of male aggressiveness manifests itself equally strongly in a population of any size, or fecundity. What about the location of the equilibria, then? An equilibrium is a point where future looks the same as the present: $x' = x$. But we also know that x' relates to x as in Eq. (2.6b) above. So we have the equilibrium condition

$$\frac{1}{2} \frac{x(1 - x + bm(1 + x))}{1 - (1 - bm)x} = x \qquad (2.7)$$

Not all values of x will satisfy this equation. Those that do, are equilibrium points: starting at an equilibrium gene frequency we predict no further change. Equation (2.7) does not readily tell us where such points are, so let us remember the rule that both sides of an equation can be multiplied by a non zero constant, yielding $x(1 - x + bm(1 + x)) = 2x(1 - (1 - bm)x)$, and then we can move everything to the left-hand side. After simplifying, we finally arrive at

$$-(1 - bm)(1 - x)x = 0 \qquad (2.8a)$$

which is the same as

$$(1 - bm)(1 - x)x = 0. \qquad (2.8b)$$

Now it is far easier to see which are the values of x that make the left-hand side of the equation vanish (which is maths jargon and means 'become zero'). Such values satisfy the equilibrium condition, and they are $x = 0$ and $x = 1$. No other equilibria are possible, unless we happen to have $bm = 1$, which also makes the left-hand side vanish. At this carefully balanced relationship between two parameters of the model, the prediction is 'no further gene frequency change' starting from *any* frequency x of the A allele. A is said to be neutral in that case.

The above calculation can also be repeated as an inequality: when is $x' > x$? We obtain the answer by replacing the $=$ sign with the $>$ sign in Eq. (2.7). Dealing with inequalities requires a little extra care, because we need to remember that multiplication with a non-zero constant may mean multiplying with either negative constant (which changes '$>$' to '$<$') or multiplying with a positive one (which does not change the direction of the inequality). Here, multiplying both sides with $2(1 - (1 - bm)x)$ is safe, because $(1 - (1 - bm)x) > 0$ whenever $x < 1/(1 - bm)$, which, in turn, must be true when $0 \leqslant x \leqslant 1$ and $bm > 0$. Shifting terms to the left while keeping tabs on the plus and minus signs leads to

$$-(1 - bm)(1 - x)x > 0 \qquad (2.9a)$$

which is the same as

$$(1 - bm)(1 - x)x < 0 \qquad (2.9b)$$

This is our criterion for the spread of the A allele. Now, when is this true? Since $(1 - x)x$ is never negative for $0 < x < 1$, the sign of this expression will depend on whether $1 - bm < 0$, or, in other words, whether $bm > 1$, which, in turn, is the same as $b > 1/m$. Isn't this nice? We have caught our salmon. "The 'aggressive' allele A will spread if the reduction of fecundity is not too large. To be exact, a fraction of fecundity larger than $b = 1/m$ should be retained, where m is the mating advantage (Fig. 2.5). For example, $m = 2$ requires $b > 0.5$. If A spreads it does so until fixation. If the fecundity drops to less than $b = 1/m$, A is selected against and a will get fixed. There are no mixed equilibria where A and a could coexist. These results are independent of fecundity B and of population size."

Beautiful – it is time to get a little excited now. Figure 2.5 plots the function $1/m$ over a range of different values of m. If the value of b exceeds $1/m$, we know that selection favours the A allele. If the value of b falls below $1/m$, the a allele is instead favoured. Now that looks a lot more professional than plotting more and more diagonal figures. Note that Fig. 2.5 is not a plot over time, nor does it predict that populations should be located on the curve. A population can exist with any combination of m and b values; therefore, it can sit anywhere in the white areas, or by chance even on the curve itself. The fate of the A allele depends crucially on which side of the curve the population resides in, since the curve $b = 1/m$ delimits areas in which A increases in frequency from those in which it decreases in frequency. If we happen to have a population in which $b = 1/m$ exactly, there is no prediction of either increasing or decreasing frequency of A: A and a are selectively neutral in that case.

It is possible, however, that you feel a little disappointed here. The rule $b > 1/m$ for the spread of A is so simple that it feels almost trivial: surely we could have guessed from the start that if the mating advantage is not outweighed by a too high fecundity cost, we will have evolution towards fixation of the A allele. Perhaps this is trivial, but triviality is (to some extent at least) in the eye of the beholder. Would you have been able to write down this equation immediately with full confidence? Better still, if you think that $b > 1/m$ is an anticlimax, try out a variant of the sexual conflict problem: intralocus conflict (Chapman *et al.* 2003). In other words, what if it is not the male genes that influence female fecundity as a result of mating, but alleles that make males successful in finding mates make a female less fecund when these same alleles are expressed in a female (i.e. regardless of who she mates with)? This is also a form of sexual conflict, although not evident in equally dramatic battles as with interlocus conflict. For example, to use an illustrative example by Bill Rice, human females could be selected to have broad hips to make childbirth less dangerous, but this could also widen the hips of males, which might be suboptimal for locomotion. Over evolutionary time, we may expect hip width to differ between the sexes, but the evolution of sex-specific gene expression is not always fast (Rhen 2000).

Applied to the present scenario, we could assume that sex-specific gene expression cannot evolve at all, and females simply suffer a fecundity decrease if they carry the A allele. All that is needed is a slight change in the equations of this chapter, but I can promise slightly more exciting results, with more diverse outcomes, than in the case of interlocus conflict. The results can be checked in Kokko and Brooks (2003). The notation in Kokko and Brooks (2003) differs somewhat from the one presented here: there F denotes female fecundity, which is standard in studies where matrix algebra is used to calculated population growth, but in the current context it is perhaps an unfortunate letter as it makes population geneticists turn their thoughts to inbreeding. The lesson? It is clearly impossible by now to standardize notation across all fields of biological study, so better acquire the belief that searching for corresponding variables between two or more scientific papers is a good alternative to crossword puzzles for keeping your mind fresh and alert.

2.2 Got interested?

Population genetic models are widely used, and introductory books include *Population Genetics: A Concise Guide* by Gillespie (1998), and *Population Genetics of Multiple Loci* by Christiansen (1999). A concise introduction can also be found in the first few chapters of Rice (2004) and in Chapters 3 and 4 of Otto and Day (2006). For early applications in the field of sexual selection, see Andersson (1982) and Hoelzer (1989). Examples of newer work in this area include a population genetic model of male and mutual mate choice (Servedio and Lande 2006), and a more system-specific model fitted to data on sexual harassment in damselflies (Svensson *et al.* 2005). Further examples can be found in many fields including life history theory (e.g. Mueller and Rose 1996), conservation ecology (Nunney and Campbell 1993), host–parasite coevolution (Peters and Lively 1999), the evolution of alarm calling (Tamachi 1987), niche construction (Laland *et al.* 1996), sex ratio theory (Wade *et al.* 2003), inbreeding depression (Tanaka 1998) and its link to dispersal evolution (Roze and Rousset 2005), and so on, all the way to an explanation of regular fluctuations in lateral dimorphism frequency in fishes (Nakajima *et al.* 2005). A keen reader will not run out of examples easily.

Sexual conflict is a hotly debated topic nowadays. A new textbook by Arnqvist and Rowe (2005) summarizes the arguments, Chapman *et al.* (2003) provide a more concise summary, and it is worth remembering that the classic reference and a must-read is Parker (1979). For a debate on whether sexual selection can cause population demise, or perhaps instead improve population viability, see Doherty *et al.* (2003), Kokko and Brooks (2003), Lorch *et al.* (2003), Morrow and Pitcher (2003), Morrow and Fricke (2004), and Radwan *et al.* (2004).

And ... back to the question phrased on p. 19. What 'additional' assumptions did we make here? Many, indeed: for example, that female or male mortality does not depend on the alleles they carry. This assumption that the cost is only seen in the fecundity makes calculations easier: had we included a mortality cost, we would have had to take into account that individuals whose alleles make them suffer higher mortality will become less prevalent in the population as they age. We have also assumed, for example, that the male mating advantage remains constant regardless of population density; see Kokko and Rankin (2006) for why this may be important. We assumed that females cannot evolve countermeasures or choose who they mate with. We assumed that we can ignore the possibility of sex-specific gene expression. We ignored drift (to begin dealing with it see, for example, Ch. 3 in Rice 2004). We in fact ignored all stochastic processes that are relevant in populations of finite size. So, is this a bad model? Not at all, it is just one that is not designed to answer questions on, say, the effects on population density (see Ch. 1).

A final question. Remember the two possibilities discussed by Crudgington and Siva-Jothy (2000) for why exactly the spiky penis exists. Does our model help to distinguish between these two possibilities: reduced remating or faster oviposition by females? Certainly not; all it is saying is that some form of mating advantage, if sufficiently strong, can counteract the damage to female reproduction when it comes to spreading in the population. Reduced remating and faster oviposition could be modelled explicitly (and indeed this task has already attracted the attention of modellers, though not in a population genetic framework (Lessells 2005)). Alternatively, one could simply try to evaluate how well these processes can be summarized with the mating advantage as modelled here. For example, can faster oviposition be viewed simply as a mating advantage for one male over other males? Perhaps yes, to some approximating extent, but certainly only if fast ovipositioning reduces the probability that a female remates before the current eggs are all fertilized and laid.

The proper route to find out what happens in reality is the usual one: observation and experimentation. Indeed, this has been done, and new results on *C. maculatus* suggest that a third explanation appears most likely: experimental females that were prevented from kicking males had to endure longer copulations and suffered lowered lifetime fecundity, as expected (Edvardsson and Tregenza 2005). But they did not remate any less than control females, or oviposit faster. This made Martin Edvardsson and Tom Tregenza suggest that the spines have evolved as a simple 'anchor' that helps to counteract kicking by the female, and that the damage to females is a simple side effect of male adaptation ('collateral harm', see Lessells 2006; Parker 2006). Does this render our model irrelevant? Again, no, the model is general enough to cover such a case too: it states that more successful mating can counteract collateral harm to females. Thus the model helps to reconfirm that the argument of Edvardsson and Tregenza is logically sound, in the same way as it could have been employed in the other explanations (ideally perhaps with m replaced with more specific paternity calculations when females remate (e.g. Johnstone and Keller 2000)). The model did not take sides at all when arguing about *why* the 'aggressive' type has higher mating success; that fact had to be empirically established.

3

Quantitative genetics

where we learn to handle a bewildering number of loci, after which a whiff of predators does not scare us at all.

In principle, the approach of the previous chapter could be extended to many loci and many different possible alleles. For example, we might want to extend the sexual conflict example to predict what happens if there are different alleles that improve mating success to a different extent, and also cause reproductive failure of variable magnitudes in females. All we need to do is to keep track of an increasing number of possible mating pairs: for example, 10 different alleles means $10 \times 10 = 100$ different combinations of males and females. Hmm . . . can be done, but is hardly tempting. Still, this complication addresses one aspect of the problem only, i.e. how allelic variation should be dealt with. It gets worse, once we consider that many phenotypic traits are influenced by a multitude of loci, plus environmental variation. Population genetic models could, in principle, be built to find out rules of inheritance and the subsequent direction of evolution in such complicated settings, but this can get very tedious. Additionally, in most cases we also operate quite blindly: even in an era where we have the genome of humans, yeast, fruit flies and whatnot sequenced, we have typically no clue how many loci are involved in determining a particular trait, let alone being able to specify the effect of each allele at every locus.

Luckily, there is quite a bulk of theory built on inheritance of polygenic traits (i.e. traits that are influenced by very many loci). All of it rests on a fine mathematical result, the *law of large numbers*. Its essence is the following. If you throw dice 100 times and calculate the mean of your score per throw, you are very likely to observe a value close to 3.5, and it is extremely unlikely that the mean is close to 6 – even though the sequence

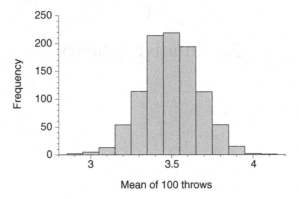

Fig. 3.1 Computerized dice throwing for 1000 replicates: in each replicate, 100 random numbers were picked with equal probability of each of the values 1, 2, 3, 4, 5 and 6, and the mean is recorded. The distribution is very close to a normal, bell-shaped distribution.

6, 6, 6, 6, ..., 6 occurs with a probability that is precisely equal to any other single sequence such as 1, 3, 2, 6, ... that has a mean closer to 3.5. The reason why the mean is likely to be close to 3.5 is that there are more sequences that produce something close to 3.5. In fact, the distribution of the mean is approximately normal (Fig. 3.1), and the difference to the exact normal distribution approaches zero when the number of throws increases (this result has a nice name, the *central limit theorem*).

What does this have to do with genetics? Imagine that each throw is the effect of a particular locus on the trait, and the phenotype (observed trait, say, body size) can be characterized as the mean effect of all loci. A locus that has the value 6 tends to produce large-bodied organisms, but because there are so many loci involved, the distribution of body sizes found in the population becomes approximately normal. The miracle of the law of large numbers is that the normal distribution is approached regardless of how weird the original distribution was: there is nothing nice and bell shaped about each side of a dice having exactly 1/6 probability, but the mean still behaves very normally (Fig. 3.1). The genetic interpretation? We don't have to know much about the genes involved and yet we can still be quite confident that a normal approximation works well. Pretty much all that is required is that there are many genes whose effect can be thought to add to each other, and none of them should have an overwhelming effect over the others.

Since we are interested in evolution, we would like to know what happens if some of the phenotypes survive or reproduce better than others. How does the underlying trait distribution change, and what is the distribution of phenotypes in the next generation? If we know how to do this, we have a very powerful tool for predicting evolutionary change.

For a moment, consider what happens if individuals produce exact clones of themselves, but the rate of offspring production differs depending on the trait in question. For example, what if large-bodied organisms are more fecund, and body size is perfectly heritable? The proportion of genotypes that produce large bodies will clearly increase in the population.

But how fast exactly? We are interested in growth rates of certain lineages. We cannot add numbers in the simplest possible (additive) way, because growth rates follow mathematics of their own. To follow a biblical expression, animals *multiply* but this also *adds* individuals to their populations. This causes some inconvenience, because it means that fitness can be expressed in two distinct ways: λ and r. Let us consider λ first. This is a multiplicative measure. For example, $\lambda = 1.2$ means that there is a 20% increase in individual numbers from one year to the next. Then, $\lambda = 1$ means no increase, and $\lambda = 2$ means doubling. How many copies of a gene will we then have in, say, 50 years, if we start from one animal?

Certainly not $50 \times \lambda$. It is once again worth getting sidetracked here, and think about money for a change. If a student is about to take a loan of 1000€ with a staggering 20% interest rate, and does not pay the bank anything back for 2 years, she does *not* owe the bank $1000 + 1000 \times 0.2 \times 2 = 1400$€. Had she thought so, the bank would be very quick to correct her and tell her that after 1 year she owes $1000 \times 1.2 = 1200$€, and after 2 years the debt is $1200 \times 1.2 = 1440$€. In other words, two 20% increases are not the same as one 40% increase. You can quickly verify that the student will end up in quite some trouble if she does not pay back for, say, 10 years: she then owes the bank more than 6000€.

But what if the student noticed the scary rate of debt accumulation quite quickly, got scared and wanted to pay back fairly immediately? What is the fair amount she should pay after 2 months? No clumsy solution such as 1.2 divided by 12 will work, because we once again have to think about the multiplicative nature of interest accumulation. But there is a clever way available. We can ask what is the *instantaneous* rate of interest accumulation that leads to the correct values of 1.2 after 1 year, 1.2×1.2 after 2 years, and so on, but also predicts values for intermediate times such as 2 months (one sixth of a year, see Fig. 3.2).

The trick is to move to a logarithmic scale of money. If $\lambda = 1.2$, then $r = \ln(\lambda) = 0.1823$ and conversely, $\lambda = e^r = \exp(0.1823) = 1.2$. Why is

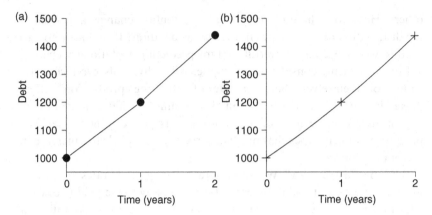

Fig. 3.2 Debt accumulation over two years (a) counting directly from one
year to the next by multiplying the current debt by 1.2 (intermediate values
are not obtained accurately) and (b) using $r = \ln(1.2) = 0.1823$, and noting
that debt at time t must equal $1000 \times \exp(rt)$. Crosses mark locations where t
is an integer (completed years). At these locations, the accumulation of debt
passes through exactly the same points as in (a). The curve in (b) is not
exactly a straight line, but it would be if the y axis was logarithmic.

$r = 0.1823$ more helpful than $\lambda = 1.2$? Because multiplicative things start
to behave additively when logarithms are taken. This, by the way, is the
same reason people used to keep slide rules in their pockets to perform
multiplications before the era of cheap pocket calculators. Within the
realm of the logarithmic scale, the growth of the debt over a year equals
the sum of 12 one-month growths. In other words, each month's debt
accumulation equals 1/12th of a year's accumulation, $0.1823/12 = 0.0152$.
Over 2 months this is $0.0152 + 0.0152 = 0.0304$. Everything behaves
nicely and additively, so the *logarithm* of debt increases linearly too: after
2 months it is $\ln(1000) + 0.0152 + 0.0152$.

But since nobody has logarithms of euros or dollars in one's wallet, one
would also like to get the answer back in 'real money'. This additive exercise
can be reversed back to the multiplicative scale by calculating
exp(quantity of interest). So, after 2 months, our student owes the bank
$\exp(\ln(1000) + 0.0304)$, in other words $1000€ \times \exp(0.0304)$, which both
give the same answer: close to 1031€. Clearly, it is much better to pay
back now than after 10 years. Note that the interest accumulated in 2
months was not quite one sixth of the 200€ that would have accumulated
in 1 year. Growth is initially slow, and the number of euros added to
the debt per time unit increases over time: later there are more euros

participating in the debt, all of them capable of growing further. Replace money with individuals, and you will see the analogy.

The linear ease with which r can be treated comes in very handy in evolutionary ecology. The scale of r values is different from that of λ. No growth, $\lambda = 1$, corresponds to $r = 0$, and a positive r indicates growth in the same way as $\lambda > 1$ indicates growth. Negative r corresponds to $\lambda < 1$.

Now consider that fitness depends on the phenotype, which in our case equals the genotype (remember that we still assume perfect heritability). A linear relationship would be particularly nice, but let's not assume such a thing while using λ. Linear relationships, for example, have a nasty tendency of becoming negative at some point, and no individual can ever produce minus two offspring. But r can take positive or negative values, no problem there: if a lineage decreases in frequency it will have a negative r. So, let's try a trick: specify a distribution of genotypes (call this z), assume that phenotypes correspond to genotypes exactly (i.e. we assume no environmental influence on the trait) and then look at the *logarithm* of fitness.

Fitness as a growth measure is often denoted by either λ or W, depending on who one talks to; λ is more often used when the emphasis is on population growth or on life history traits, while W is often used in formal quantitative genetic analyses. Either way, the results will be the same. The natural logarithm, $\ln W$, must then equal the growth rate r of this genotype: according to our definition $r = \ln W$ when $\lambda = W$.

Then, let r depend linearly on the phenotype. The phenotype equals z since we assume (so far) that the phenotype reflects the genotype exactly. The linear dependence can be expressed as $r(z) = a(z - b)$, which here[1] means that a is multiplied by $(z - b)$. This is perhaps not the most familiar way to express linearity: perhaps more commonly, one finds expressions of the form 'coefficient $\cdot z$ + constant', where z is the variable of interest. But $a(z - b)$ is linear too. It can also be written as $az + (-ab)$, so the matter is clear after noting that the 'coefficient' now corresponds to a and the 'constant' takes the value $-ab$. The reason for writing linearity as $a(z - b)$ here is one of convenience. Both a and b are easily interpretable

---

[1] In another context, the notation $a(z - b)$ might mean that a is a function of something (say x), and x right now has the value $z - b$. One does often get the feeling that mathematics would be easier to learn if we scrapped all current notation practices and started from scratch. True, but difficult to achieve, and the problem is not unique to mathematics: German spelling was simplified in the 1996 *Rechtschreibreform*, but this met so much resistance that some major newspapers and periodicals never switched to the new system, or used it only briefly. There have been repeated attempts to do the same with the much more chaotic English spelling (e.g. Benjamin Franklin invented a new alphabet for this purpose), but these have always failed.

now: a is the slope of the relationship (i.e. how steeply fitness depends on, say, body size) and b is the precise value that separates the region in which $r(z) < 0$, namely $z < b$, from those in which $r(z) > 0$, namely $z > b$.

How to find the number of individuals in the next generation having a particular value of z? If there are $f_0(z)$ individuals now in generation 0, and each individual multiplies according to its own value of $r(z)$, then the number of individuals in generation 1 must equal $f_1(z) = f_0(z)\exp(r(z))$. Individuals that have large body sizes become more common as they have positive $r(z)$, while small-bodied individuals become rarer as their $r(z)$ is negative.[2] The net effect is that the distribution of z values as a whole shifts to the right (Fig. 3.3 and Box 3.1). Conveniently for our argument, the distribution also stays normal,[3] and the magnitude of the shift appears to follow simple rules: it is proportional to the slope a (compare Fig. 3.3a with Fig. 3.3b), and to the variance of the normal distribution of phenotypes (compare Fig. 3.3a with Fig. 3.3c). This is natural selection in action. The linear behaviour of r means that the mean will shift exactly the same distance in every generation; the cumulative change can be added up as the sum of changes in each generation.

Box 3.1

How to create Fig. 3.3 in MATLAB. The program below should first be saved as the file `distr_shift.m`. Then, calling this function in the following way

`shift = distr_shift(0.1, 15, 1, 15)`

will produce Fig. 3.3a, and also gives the answer that shift of the mean equals precisely 0.1.

```
function [shift, f0, f1, z] = distr_shift(variance,    ⎫ one
    initial_mean,a,b)                                   ⎬ line
% function [shift,f0,f1,z] = distr_shift(variance,
```

---

[2] A careful reader might notice that frequencies at $z = 15$ in Fig. 3.3 shift downwards, even though at $z = 15$ we have $r = 0$, which predicts no change. This is because Fig. 3.3 plots frequency distributions rather than actual numbers of individuals, which necessitates a normalizing procedure: all frequencies must sum up to 1 (see Box 3.1). This represses everyone's growth rate by the same (logarithmic) amount and makes only large positive values of r indicate net growth. If we plotted the number of individuals and not their frequency, we could skip the normalizing. We then obtain a distribution that keeps a constant value at $z = 15$, keeps a normal shape, and its mean shifts to the right by the same distance as in Fig. 3.3. This can only be achieved when the total area under the curve keeps growing, which indicates a growing population. It is easy to modify Box 3.1 to verify this.

[3] If you want to prove this, the key is to have a look at the shape of the normal distribution and notice that it is the exponent of a parabolic shape. So the shape should stay like this after being multiplied by $\exp(a(z - b))$. Then, use the rule $e^x e^y = e^{x+y}$ and see what happens.

```
% initial_mean,a,b)
% Graphical explanation for the shifting
% distributions,
% given the additive genetic variance, the initial
% mean of a distribution,
% and a and b of the function r(z) = a(z-b).

% Outputs are the magnitude of the shift, and the old
% and new distributions f0 and f1, which can be
% plotted against z.

z = linspace(13, 17, 201); % we create 201 different
% values of z that lie between 13 and 17 (note that the
% function is not very general as we assume the
% initial mean will fall between these values)

% the following is the density function of the normal
% distribution
f = 1/(2*sqrt(variance))*exp(-0.5*
    ((z-initial_mean)/(sqrt(variance)))).^2);

% Then normalize it, which means ensuring that values
% sum up to 1
% - they might not initially because we only have a
% selection of discrete values of z, not 'all
% possible' values
f0 = f/sum(f);
r = a*(z-b);
f1 = f0.*exp(r); % the new distribution
f1 = f1./sum(f1); % normalize f1 as well
plot(z,f0,z,f1)
% calculate the new mean as a weighted sum of values of z
new_mean = sum(f1.*z)
shift = new_mean-initial_mean;
```

}one line (annotation for the `f = ...` statement)

Of course, Fig. 3.3 is a simplification of a more general setting in nature. Traits are very rarely perfectly heritable: large body size could, for example, result from being lucky enough to locate a good resource patch early in life. This makes it obviously harder to compute trait distributions. Indeed, at first sight, it appears horrendously complicated. Consider a hypothetical example, where the largest three individuals get to reproduce, and everyone else fails (e.g. because the largest win ownership

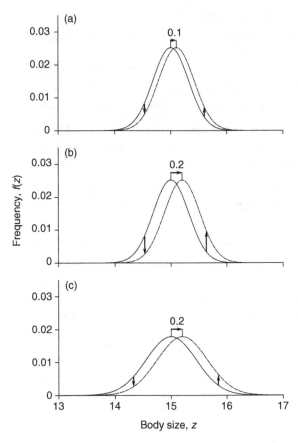

Fig. 3.3 Shifting phenotypic traits. Body size z has initially mean value of 15, and it is normally distributed around this value (solid line). The dashed line gives the predicted distribution after one generation of selection, when $r(z) = a(z - b)$ (see text for explanation). When $b = 15$, large phenotypes with $z > 15$ become more common, as indicated with arrows pointing upwards. Small phenotypes with $z < 15$ become rarer, indicated with downwards arrows. The result is that the whole distribution shifts to the right. The change in the mean (length of arrow pointing to the right) is (a) $\Delta \bar{z} = 0.1$, when $h^2 = 1$, $a = 1$ and $\sigma_z^2 = 0.1$, (b) $\Delta \bar{z} = 0.2$, when $h^2 = 1$, $a = 2$ and $\sigma_z^2 = 0.1$ and (c) $\Delta \bar{z} = 0.2$, when $h^2 = 1$, $a = 1$ and $\sigma_z^2 = 0.2$.

of the local resource). It is simple enough to deduce that in Fig. 3.4a we expect no shift in the distribution of genotypes, no matter how eagerly one selects for the large-bodied organisms: there is no heritability in the trait. But exactly how much does the distribution shift in the case of Fig. 3.4b, where there is a clear relationship between genotype and phenotype, but some environmentally determined variation appears to

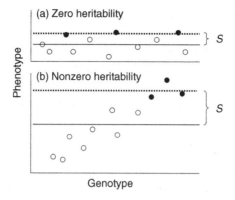

Fig. 3.4 Examples of selection where, in both (a) and (b), only largest phenotypes (e.g., largest bodied organisms) can reproduce: these are marked with filled dots. The selection differential S indicates the difference between the average phenotypes of parents compared with an average member of the parental generation (including those who do not reproduce). In (a), there is zero heritability: there are different phenotypes in the population, but the phenotype does not depend on the genotype in any predictable manner. Consequently, the genotypes of parents (filled dots) are a random subsample of the population, and we expect no evolutionary change despite $S > 0$. In (b), there is heritability: only genotypes near the right end of the distribution will be present in future generations; therefore, we predict an evolutionary response to selection S. Note that the fact that S is larger in (b) than in (a) strengthens the response but is not a sufficient explanation for a complete lack of evolutionary change in (a); heritability has its own, independent effect (Eq. 3.1).

cause some noise too? Perhaps we should track every single possible combination of alleles, together with any values of the environmental noise that cause deviations between exact expression of genotypes at the phenotypic level ... this appears very tedious ...

... but luckily it is not needed at all. Very kindly, theoreticians have gone through this problem in detail and have shown that the answer can be summarized in a very, very simple and neat equation. So, even though it is good to keep in mind how and why the vertical change of the distributions leads to a horizontal shift (Fig. 3.3), in practice we will never again have to look at the details of this process. Instead, we can simply trust that a mightily compact expression,

$$\Delta \bar{z} = h^2 S \tag{3.1}$$

deals with everything we need. Here $\Delta \bar{z}$ is the shift in the trait mean over one generation (the line over z indicates that we're talking about a mean value), and h^2 is the heritability of the trait, which can be written as σ_A^2/σ_z^2: the ratio of additive over phenotypic variance. This ratio reflects the fact that not all observed variation is inherited. The phenotypic variance could be high, for example, because some individuals are better fed than others and, therefore, larger. Selecting these individuals to breed would not make a difference at all to evolutionary change – the real 'fuel' of the evolutionary process is the additive variance present among the individuals' genotypes.

The ratio that defines heritability is zero in Fig. 3.4a, while it is quite high in Fig. 3.4b. The term S is the selection differential, in other words a measure of how much parents of the new generation have trait values that differ from the population mean. In Fig. 3.4, S is the height of the dotted line compared with the population mean. Equation (3.1) is present in all standard textbooks on evolutionary genetics.

The example of Fig. 3.3 was made easy by assuming perfect heritability ($h^2 = 1$), so any differences in the responses by the trait distribution must have been caused by changes in S. We observed that the trait mean moved more if the trait distribution was broader; also, it moved more if fitness depended on body size more steeply: that is, $\ln W$ (also denoted r) was strongly dependent on z. These observations are not isolated examples but confirm a general expression of how S depends on phenotypic variance and on fitness:[4]

$$S = \sigma_z^2 \frac{\partial \ln W}{\partial z} \qquad (3.2)$$

The term σ_z^2 is the observed (phenotypic) variance in body size, and the expression that follows tells us how fitness depends on body size. What is this odd squiggle, $\partial/\partial z$? Scary as it might look, it is quite a tame creature that answers to the name *partial derivative*. It is there simply to tell you 'how much does $\ln W$ change, if we increase z a little bit'.[5] In other words, this is the slope of r against z, which in our example (Fig. 3.3) is the same as the parameter a. In more biological terms still, if we compare a

---

[4] All of this holds given some assumptions. For example, selection should be *weak*, which among other things means that there are no over-rapid changes in the way that r depends on z causing it to start distorting genotypic distributions in funny ways: loosely speaking, we assume that as one end of the distribution thickens, the other end becomes thinner at a similar enough pace.

[5] The word 'partial' appears here because W can also change if other parameters are increased, but we choose to view the change along one axis only. In the barnacle example of Section 3.1, W depends, for example, on how dangerous predators are: more danger means that α_1 and β_1 decrease, and one could examine, for example, $\partial W/\partial \alpha_1$ to gain insight on that aspect of the system.

medium-size animal to one that is slightly bigger, exactly how different
are their fitness values?

Our numerical calculation produced a shift of $\Delta\bar{z} = 0.1$ in Fig. 3.3a.
In that example, $h^2 = 1$, $\sigma_z^2 = 0.1$ and $\partial\ln W/\partial z = a = 1$. This means that
we indeed predict a shift of $\Delta\bar{z} = 0.1$: from Eqs. (3.1) and (3.2) together,
$1 \times 0.1 \times 1 = 0.1$. It all fits happily together, and we can also now verify
that our predictions are correct in the examples shown in Fig. 3.3 too.

Often, S is written in the form $\sigma_z^2(1/W)(\partial W/\partial z)$. Why is this the same
as $\sigma_z^2(\partial\ln W/\partial z)$? This follows from rules of taking derivatives, when we
remember that the derivative of $\ln W$ is $1/W$. A derivative is a measure of
how much something changes. So, the fact that $\ln W$ has the derivative
$1/W$ tells us several useful things. The increase is very fast if W is small
since then $1/W$ is a big positive number. It is much less quick when W is
large, since $1/W$ is then a small positive number; but it always increases,
since $1/W$ is always greater than zero. (More on derivatives can be found
in Ch. 4, see Fig. 4.1.)

There is nothing particularly exciting about these two forms of S, but it
is often quite useful to have these two ways to express it, because one of
them might be easier to use in a particular scenario. Indeed, it is the form
containing the $1/W$ that we will use in the examples of this chapter. It
follows that

$$\Delta\bar{z} = h^2 S = \frac{\sigma_A^2}{\sigma_z^2}\left(\sigma_z^2 \frac{1}{W}\frac{\partial W}{\partial z}\right) = \sigma_A^2 \frac{1}{W}\frac{\partial W}{\partial z} \tag{3.3}$$

Phew! The theory is now there, so let us now have a look at a real-life
example, to see what the derivatives translate to in the big wide ocean.

3.1 Bent barnacles

Figure 3.5 shows two morphs of the acorn barnacle *Chthamalus anisopoma*: one looks like a normal barnacle with its conical shell, while the
other shell is curiously bent. The bent morph forms when there are predatory snails *Acanthina angelica* present (Lively 1986a,b; Hazel *et al.*
1990); these morphs are more resistant to attacks by the predator, but
developing this form seems to come at a cost of reduced fecundity, and the
bent morph grows slower too (Lively 1986a).

Highly intriguingly, the conical and bent morphs do not seem to be
genetically determined alternatives, but barnacles are able to detect the
presence of the snails, presumably through some chemical cues. Where

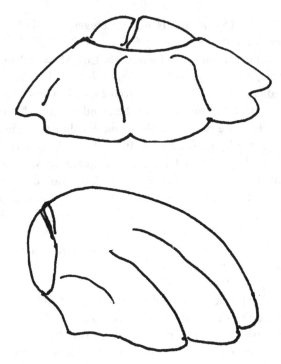

Fig. 3.5 The normal and the bent morph of the acorn barnacle. For more detail on the morphology see Lively (1986a,b).

the predators are absent, virtually all barnacles develop as the 'normal' morph, but in the presence of predators 20–48% develop the bent shape (Lively 1986a; Hazel *et al.* 1990). This is an example of a *reaction norm*: a life history response to an environmentally determined cue. The genes present in barnacles do not determine the morph directly; instead, they control the reaction norm that dictates how to respond to a given environmental situation. In other words, the same set of genes can lead to different phenotypes depending on the environment the organism develops in. And conversely, the same environment can lead to different phenotypes, depending on the alleles that an individual possesses.

But if the bent morph is so much better where the predator is present, why don't barnacles develop a reaction norm that makes them all become bent in the presence of the predator? The modelling exercise here is a simplified version of the polygenic (i.e. quantitative genetic) model of Hazel *et al.* (1990), which is (partly) devoted to answering this question.

So, how should we begin to model this question? Clearly, at the phenotypic level there are two morphs. One is defended, the other undefended. Let us label these as A and B, respectively. Then, predators could be either present or absent. The fitness W of an individual can take four different values:

- defended (A), predator not there: α_0
- defended (A), predator is there: α_1
- undefended (B), predator not there: β_0
- undefended (B), predator is there: β_1.

How do we know what values to assign? We don't, of course; we are only assuming that there must be some value that describes the fitness of an undefended individual in the presence of a predator – and given this unfortunate situation, this value β_1 is probably quite low. This is a start, but now we must begin with the modelling of a reaction norm. The barnacles must respond to some sort of cue that correlates with the presence of a predator. However, barnacles probably do not have as accurate information about this as the researcher studying the system. They have to rely on an environmental cue, which we could denote by x. Where x (e.g. the concentration of some odour) is high, it is quite likely that predators really are there; if x is low, it is unlikely that predators are there. However, a variety of factors mean that a predator *can* be encountered even if x is fairly low, or can be absent even if x is high: for example, odours flow with water, the machinery of barnacles to detect olfactory cues is not perfect, snails can move or die, and so on. For this reason, it is impossible to state that below a certain value of x there are no predators, or that above this value there are definitely predators around. Instead, a far more likely scenario is the following. The probability that the site has predators, which we denote by y, is an increasing function of the environmental cue x. The simplest possible choice is $y = x$: the probability that a predator is there increases linearly with the environmental cue. For this to make sense, we should measure the cue x on a scale that makes it vary between 0 and 1, for example, 1 could be the maximum concentration of an odour cue when predators are abundantly present.

The problem faced by a developing barnacle is that it must develop into one morph or the other, even if x equals, say, 0.3. With this value of x, it cannot be said for certain if predators are there or not. Rather, there are predators with 30% probability since $y = x$. What kind of reaction norm is favoured: one that results in a normal or a bent shape in this particular situation? In other words, how large should x be before the barnacle should 'take the threat seriously' and develop as the bent morph?

Now, we shall make the assumption that the *switchpoint* of the reaction norm is a polygenic trait, and we denote this by z. A switchpoint with value z means that if the barnacle perceives that the environmental cue x has a value that exceeds z, then the barnacle develops into the bent morph; otherwise, it develops the normal shape. (Do we actually know that the switchpoint z is a character controlled by many loci? Not really... but since almost any complex trait is, we can probably make this assumption quite safely.) The mean of the distribution is given by $\bar{z}$, and its variance by σ_z^2, of which σ_A^2 is additive genetic variation. But how do we know the values of these quantities before we set evolution on the loose? Of course, we have no idea. The point is to examine many different values, to gain insight on all the possible worlds.

From Eq. (3.3), we immediately see what the goal is: to derive $\Delta\bar{z}$ and therefore to be able to conclude where evolution will lead us. The expression contains σ_A^2; fine, this is just a parameter that we can vary and record its effects. But W and $\partial W/\partial z$ require a little more attention here.

Consider an individual with a current switchpoint that equals z. What do we need to know to find out its fitness? Two things: how often it ends up being defended or undefended (A or B) given the current environment, and how often it actually encounters predators when being phenotypically A or B in the current environment.

Here, we notice that there is something we have not thought about defining yet. How often do individuals find themselves in situations with different strengths of the environmental cue x (and hence, since $y = x$, how often are predators really there)? Any distribution of course is possible here: if predators and hence the environmental cues are truly rare, then low values of x are much more prevalent than large values. But, for the sake of simplicity, let us assume that every value of x between 0 and 1 is equally likely: in other words, a uniform distribution. This simplifies the following analysis, as it allows for a nice geometrical determination of fitness W.

Figure 3.6 exemplifies the argument for $z = 0.4$. If the strength of the environmental cue falls below z, the individual develops as the undefended morph B. Question one: what is the probability that this happens? And, question two: given that it happens, what is the probability that undefended development is a mistake (i.e. the site in fact has predators)? Answering question one, the probability must be 0.4 (or more generally, z), given that every cue from 0 to 1 is equally likely to occur. What about question two? Again, every cue value below z is equally likely, and since values of x are associated with a linearly increasing probability y that the

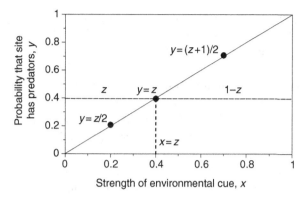

Fig. 3.6 A graphical arguments for deriving fitness, W, for normal and bent morphs of barnacles. See text for details.

site has predators, the overall probability for a predatory site, for individuals that develop as morph B, is the average between $y = 0$ and $y = z$, which is $z/2$.

If, however, the strength of the cue exceeds[6] the switchpoint value z (which happens with probability $1 - z$), then the true probability of a predator being present is the mean of $y = z$ and $y = 1$, which is $(z+1)/2$ (Fig. 3.6). So, to recap:

- the probability that $x < z$ is z
- the probability that, given $x < z$, the site has predators is $z/2$
- the probability that $x > z$ is $1 - z$
- the probability that, given $x > z$, the site has predators is $(1 + z)/2$.

So, the fitness of our individual who uses switchpoint z equals

$$W = z\left(\frac{z}{2}\beta_1 + \left(1 - \frac{z}{2}\right)\beta_0\right) + (1 - z)\left(\frac{1+z}{2}\alpha_1 + \left(\frac{1-z}{2}\right)\alpha_0\right) \quad (3.4)$$

Whoa! Neat equation, and not as complicated as it first looks. All it does is to keep track of the probability of every possible scenario and multiply it with the associated fitness consequence. The first z spells out the probability that the cue x falls below the switchpoint z. If that is the case, then we also know that the site has predators with probability $z/2$, which yields fitness β_1 since our poor barnacle is not defended, and with the

---

[6] To be exact, the case where x precisely equals z should be included in one or the other case too, so that one of them reads $x < z$ and the other $x \geq z$. But this has no influence on solutions when we are interested in continuous distributions for x and z.

remaining probability $1 - z/2$ there truly are no predators, so fitness equals β_0. The terms with α_1 and α_0 are similarly calculated, and these occur when the cue exceeded the switchpoint (probability $1 - z$).

Next, we must derive $\partial W/\partial z$. Equation (3.4) shows how W depends on z. A derivative is a description of how fast W changes if we change z. Any standard textbook of calculus, or nowadays a mathematics program, contains many rules on how to find the derivative for a specific function. For example, the derivative of z^2 is $2z$, which captures the information that the function z^2 decreases when $z < 0$ (note that $2z < 0$ whenever $z < 0$) but increases when $z > 0$, and the rate of increase is larger when z is large. Finding the derivative of Eq. (3.4) requires checking a few of such derivation rules, but once this is done, the result is $z(\beta_1 - \alpha_1) + (1-z)(\beta_0 - \alpha_0)$. Not bad: expressed in terms of the selection differential S, this means

$$S = \sigma_z^2 \frac{1}{W} \frac{\partial W}{\partial z} = \sigma_z^2 \frac{1}{W}(\bar{z}(\beta_1 - \alpha_1) + (1 - \bar{z})(\beta_0 - \alpha_0)) \qquad (3.5)$$

Why did we switch from using z to $\bar{z}$ here? Because we are evaluating $\partial W/\partial z$ at values of z that are close to the population-wide mean, which equals $\bar{z}$. The change in this mean over a generation will be (from Eq. 3.3)

$$\Delta\bar{z} = h^2 S = \frac{\sigma_A^2}{W}(\bar{z}(\beta_1 - \alpha_1) + (1 - \bar{z})(\beta_0 - \alpha_0)) \qquad (3.6)$$

The first term, σ_A^2/W, states that the speed of evolution is proportional to the amount of additive genetic variance. Fine. Since this term cannot ever be negative, the more interesting term is undoubtedly the latter, $\bar{z}(\beta_1 - \alpha_1) + (1 - \bar{z})(\beta_0 - \alpha_0)$. This can be positive or negative, which means that it can influence the *direction* in which the trait mean moves.

Let us pause for a moment and think about biologically sensible values for β_1, α_1, β_0 and α_0. Sensible values should clearly have $\alpha_0 < \beta_0$, because defence is costly when it is pointless. Also, we should have $\alpha_1 > \beta_1$, because it is good to be defended where predators really are present. Probably it also makes sense to assume $\alpha_0 > \alpha_1$ and $\beta_0 > \beta_1$, because the predator's presence is harmful whether one is defended or not. Defence in the case of morph A is probably not such a great invention that it would improve fitness over what the defended morph can achieve in the absence of the predator.

In total, a sequence of values that satisfies these criteria are $\beta_0 > \alpha_0 > \alpha_1 > \beta_1$. To give this sequence some biological meaning, one can state that the difference $\beta_0 - \alpha_0$ measures the costs of defence. This difference reflects how much fitter undefended morphs are when there are no predators present. The term $\alpha_0 - \alpha_1$ gives an indication of the severity of predation in the case of being defended: a large difference indicates that the presence of predators matters to defended morphs too. Finally, $\alpha_1 - \beta_1$ indicates how efficient defence is relative to the undefended morph.

Before turning to examples, it is good to do some salmon catching (see p. 32): find analytical solutions. Can we predict evolutionary endpoints, that is, equilibrium values where no further evolutionary change is predicted? In other words, when is $\Delta\bar{z} = 0$?

Assuming that we have chosen sensible values for α and β, so that $W > 0$, it follows from Eq. (3.6) that $\Delta\bar{z} = 0$ can only be achieved when $\partial W/\partial z = 0$. In other words, we would like to know the value of $\bar{z}$ (let us denote the equilibrium by $\bar{z}*$) for which

$$\bar{z}*(\beta_1 - \alpha_1) + (1 - \bar{z}*)(\beta_0 - \alpha_0) = 0 \qquad (3.7a)$$

Solving this yields

$$\bar{z}* = \frac{\beta_0 - \alpha_0}{(\beta_0 - \alpha_0) + (\alpha_1 - \beta_1)} \qquad (3.7b)$$

The solution has been arranged in such a way that the biologically relevant relationships are easy to see. The equilibrium switchpoint moves towards higher values if the costs of defence increase ($\beta_0 - \alpha_0$ is high), and towards lower values if defence is very efficient ($\alpha_1 - \beta_1$ is high). Remember that a high value of the switchpoint means that the defended morph becomes rarer: a larger value of the environmental cue (e.g. odour) is required before the switch occurs. Interestingly, the severity of predation in the case of being defended, indicated by $\alpha_0 - \alpha_1$, does not influence the solutions at all if $\alpha_0 - \alpha_1$ is changed in such a way that $\beta_0 - \alpha_0$ and $\alpha_1 - \beta_1$ are kept constant.

Figure 3.7 shows some examples (see Box 3.2 for how to create this figure). The three scenarios in the figure cover the situations:

(a) Defence is highly costly but very efficient, and the severity of predation drops close to 0 if defended[7]: $\beta_0 \gg \alpha_0 \approx \alpha_1 \gg \beta_1$. At equilibrium, the switchpoint is close to 0.5.

(b) Defence is not too costly, but not very efficient either. Predation is never very severe, so that all fitness values are roughly similar,

---

[7] Mathematicians have a delightful tendency to create precise notation for everything, including the description of inaccuracies. Here $\gg$ means 'much larger than', and $\approx$ means 'roughly equal to'.

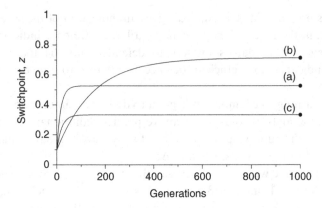

Fig. 3.7 The evolutionary process, exemplified for three sets of values for $\beta_0 > \alpha_0 > \alpha_1 > \beta_1$. For illustrative purposes, the initial distribution of z has in all cases the same mean $\bar{z} = 0.1$, and the additive genetic variance $\sigma_A^2 = 0.1$. Dots at the end of 1000 generations indicate the location of the predicted equilibrium, $\bar{z}^*$. (a) $\beta_0 = 2$, $\alpha_0 = 1.5$, $\alpha_1 = 1.45$ and $\beta_1 = 1$ leads to an equilibrium value of $\bar{z}^* = 0.5263$. (b) $\beta_0 = 1.1$, $\alpha_0 = 1.05$, $\alpha_1 = 1.02$ and $\beta_1 = 1$ leads to $\bar{z}^* = 0.7135$. (c) $\beta_0 = 4.5$, $\alpha_0 = 4$, $\alpha_1 = 2$ and $\beta_1 = 1$ leads to $\bar{z}^* = 0.3333$. Note that in all of the examples, the switchpoint evolves upwards, meaning that evolution leads to antipredatory defence being manifest in *fewer* individuals over time.

$\beta_0 \approx \alpha_0 \approx \alpha_1 \approx \beta_1$. Evolution is quite slow, but eventually a high value of $\bar{z}^*$ is reached such that few individuals are defended.

(c) Defence is moderately costly, and predation risk is high. Defence is efficient in the sense that being defended doubles fitness in the presence of predators. However, even defended prey get eaten quite often: the difference between α_0 and α_1 remains large. This example leads to $z^* = 0.3333$, and most individuals develop into the defended morph.

Box 3.2

In MATLAB, figures such as Fig. 3.7 are produced like this. Firstly, the program below is saved as the file `barnacle.m`. Then, for example, Fig. 3.7a is produced by calling `barnacle([2 1.5 1.45 1],.1)`. The initial help lines (marked with %) are retrieved in MATLAB if one types, amusingly enough, `help barnacle` in the command window.

```
function [z, equilibrium]=barnacle(params,            ⎫ one
    additive_var)                                      ⎬ line
% function [z, equilibrium]=barnacle (params,         ⎭
% additive_var)
```

```
% 'params' should contain b0, a0, a1 and b1
% ... arranged in this weird order to make it easy to
% remember that the sequence of values
% should be decreasing
% additive_var is the amount of additive genetic
% variation.
% The output includes the mean value of z over time,
% and the predicted equilibrium value (a0-b0)/
% (b1-b0+a0-a1).
z=0.1;
b0=params(1);
a0=params(2);
a1=params(3);
b1=params(4);
maxt=1000;
% the number of generations could of course
% have been made an argument of this
% function rather than stating it here
for t = 1:maxt-1
  % first Eq. (3.4)
  W = z(t)*(z(t)/2*b1 + (1-z(t)/2)*b0) + (1-z(t))*...
    ((1 + z(t))/2*a1 + (1-t))/2*a0);
  % then Eq. (3.6)
  deltaz = additive_var/W*(z(t)*(b1-a1) + ...
    (1-z(t))*(b0-a0));
  % the new z is the old one plus the change that
  % occurred
  z(t + 1) = z(t) + deltaz;
end;
plot(z); xlabel('Generations');
ylabel ('Switchpoint')
equilibrium = (a0-b0)/(b1-b0 + a0-a1)
```

Note that the last z will be the thousandth one, even though t ran only from 1 to maxt-1, i.e. to 999. This is because the loop calculates $z(t+1)$, i.e. up to $z(999+1)$.

In all our cases, evolution proceeded towards the predicted equilibrium value, but evolution is clearly slowest when all fitness values are close to each other. This makes good intuitive sense, as does the fact that the lower the heritability, the slower the evolutionary response even if the selection differential S stays the same – try this out!

Even though we did not vary the presence of predators per se (we always assumed the same uniform distribution), or the order $\beta_0 > \alpha_0 > \alpha_1 > \beta_1$, we found widely different conclusions regarding the switchpoint, and hence very different predicted frequencies of the two morphs. Life is not that simple for a barnacle: no matter how unpleasant (more scientifically, how detrimental for fitness) being predated upon is, the first whiff of a predator should not produce a defended morph if, for example, predation is not that frequent even in sites where predators are present. This would imply that undefended morphs are relatively safe too. Since the only benefit of α_1 over β_1 is more efficient defence, then clearly α_1 cannot exceed too much beyond β_1 in this particular example. From Eq. (3.7b) we then see that a high value of $\bar{z}\,^*$ is predicted; that is, moderate cue values x will not exceed the switchpoint, and the bent morph will not be produced.

In general, animals often seem to pay less attention to predator avoidance than the emphasis of nature documentaries on trophic chains would suggest to be sensible. Differences in reproductive success can often be more important than differences in survival, which helps to explain why individuals often just get on with their lives despite the continuous threat of predation.

3.2 Got interested?

Here, we made the simplifying assumption that all cues occur with equal frequency. Hazel *et al.* (1990) show how more general results can be provided: the idea is to integrate over the frequencies of different environmental cues. What this means is that commonly encountered scenarios are given more weight in fitness calculations. They make the point that predation by snails is indeed quite a rare event, because the spatial distribution of snails is concentrated around rocky crevices. This helps to explain why so few barnacles develop the bent morph, although additional factors can be relevant too, such as gene flow from populations in which the snail is absent. For a review of antipredatory behaviour in general, see Lima and Dill (1990), and for different ways to model phenotypic plasticity, see Pigliucci (2005).

Our example was a first step in quantitative genetic modelling, considering one trait (the switchpoint) only. The real strength of quantitative genetics is that it can also be analysed with almost equal ease when there are several traits coevolving. In this case, it is not sufficient to consider one value of the additive genetic variance only. Traits very often *covary*, and in this case selecting for one trait – say, higher fecundity – also produces an evolutionary response in the other trait, for example larger body size. Negative covariation is also possible, for instance when there are life history trade-offs: selection for increased developmental speed could yield individuals with reduced fecundity. The signs of the covariation can change once again, if developmental speed, fecundity and body size all covary with each other (Roff 2000). Yet another way to find covariation is when mating behaviour produces genetic linkage. An often used example is one of female preferences for males that have extreme phenotypes, such as long tails: a genetic correlation arises because females with strong preferences tend to mate with males with particularly long tails (for reviews, see Bakker (1999), Mead and Arnold (2004) and Kokko *et al.* (2006a)).

Dealing with multiple traits requires replacing the additive genetic variance of this chapter with a so-called G-matrix, which can be thought of as a tabulated array of variances and covariances between traits. The subsequent analysis otherwise proceeds exactly as in the example above, but one has to learn a little bit of matrix algebra to deal with it. Roff (1997), Ch. 7 in Rice (2004), and Ch. 8 in McElreath and Boyd (2007) give basic instructions. Both Rice (2004) and McElreath and Boyd (2007) also include an appendix of basic rules for manipulating derivatives. Any calculus textbook will have these too. For examples that use quantitative genetic modelling, see Lande (1982) on life histories, Kirkpatrick and Barton (1997) on the evolution of species ranges, Ronce and Kirkpatrick (2001) on the evolution of ecological specialization, Kölliker *et al.* (2005) on parent–offspring interactions when parents provide care, and Baskett *et al.* (2005) on the evolution of size at maturation in fish populations that live in areas with marine reserves (which could potentially help in avoiding the problem that fisheries generate an evolutionary response to mature earlier and at a smaller size (Olsen *et al.*, 2004)). Lynch and Walsh (1998) offer a very thorough presentation of the many uses of this method.

Much of quantitative genetics deals with empirical datasets, looking for ways to partition the phenotypic variance into its components to get at evolutionary predictions. Models take these components as given and then derive evolutionary trajectories (our plot of the switchpoint's evolution over time is an example of such a trajectory). When deriving our results, we assumed that the amount of additive genetic variance is not depleted over time, which in the more general case extends to the statement that the G-matrix stays constant. For a discussion of the validity of this assumption, see Steppan *et al.* (2002), Jones *et al.* (2003) and Björklund (2004). Pigliucci and Schlichting (1997) have provided a highly readable discussion of this problem, and they also discuss other simplifications inherent in quantitative genetics modelling. More recently, Pigliucci (2006) has provided a sharp critique of the method in general; while his points raised relate to statistical analysis of populations, the issues apply to modelling conclusions too – so have a look and form your own opinion. The 'pros and cons' of using quantitative genetics models have often been compared in relation to approaches that focus on phenotypes. These approaches are introduced in the following chapters (see Chs. 4 and 6 for references concerning this debate). To mention one example that is particularly relevant to the current chapter, Hazel *et al.* (2004) have compared phenotypic and quantitative genetic approaches in the particular example of barnacle morphs, and find – unsurprisingly, but reassuringly – that both give similar answers when dealing with similar constraints.

4
Optimization methods

where spiders get quite exhausted, and the author confesses an
embarrassing mistake from the distant past

In the previous chapter, the evolutionary endpoint was one where the
selection differential S 'vanished', which is jargon and means that it
became zero: no further selection is then operating. The value of z that
made S vanish was the one that made the derivative of fitness equal zero.
Why did we calculate derivatives of fitness there? The derivative is
something about how steeply a function increases (positive derivative) or
decreases (negative derivative). This means that derivatives are very
handy for finding out where bigger values of something can be found: just
follow the uphill slope until ... it turns negative, oops, now go no further. If we are dealing with a nice smooth function (there is some
mathematical terminology for this, but just imagine something that bends
gently without sharp edges), the 'uphill' part of the slope means a positive
derivative (Fig. 4.1), and the point where uphill turns into downhill is
where the derivative passes through zero on its way from positive to
negative. In other words: where the derivative is zero, we have a good
candidate for a local maximum of the function we were interested in.[1]

So, the point where the selection differential S reached the value zero,
also happened to maximize W. This is not a coincidence since S was
indeed proportional to the partial derivative $\partial W/\partial z$. This is a deeply
significant fact: it means that if we were simply interested in the value of z

---

[1] Why only a candidate? Because if the derivative equals zero, the function could also have
reached, for example, a local minimum (Fig. 4.1), or a moment of hesitation before
increasing again, or a line that keeps flat for a while before doing that. Investigating
Fig. 4.1 for a moment should give a clue as to what is different between those arrows
that predict a local minimum, and those that predict a local maximum. We will return to
this (p. 73).

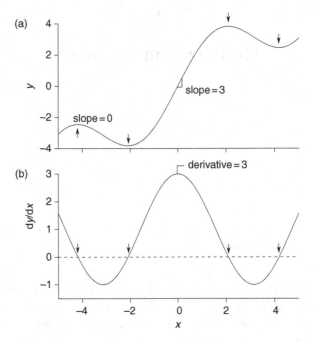

Fig. 4.1 An example of a function (a), and its derivative (b). The function here is $y(x) = x + 2\sin(x)$. Rules of derivation yield $dy/dx = 1 + 2\cos(x)$. For example, when $x = 0$, we have $y(x) = 0$, and $dy/dx = 3$. This means that, at this point, the function has the value zero but is increasing three units for every unit of increase in x; in other words, its slope is 3, and this is indeed the value of dy/dx: $1 + 2\cos(0) = 3$. Conversely, the points where the derivative is zero (arrows in b) correspond to points where the slope of $y(x)$ is zero (arrows in a). In other words, the function $x + 2\sin(x)$ has a local maximum or a minimum exactly where $1 + 2\cos(x) = 0$.

reached at equilibrium (and not in predicting, for example, how many generations this might take), we could have chosen to ignore all the underlying genetic assumptions, and simply ask, what value of z maximizes fitness. This is where we expect our barnacles to be. We would have got precisely the same answer as in Ch. 3, but in a 'shortcut' way. This is an example of a *phenotypic approach* to modelling: forget everything about genetics since it is quite likely to be unknown anyway[2] and simply assume that enough evolutionary time has elapsed that natural selection has moulded phenotypes to answer the current environmental

[2] See Rosales (2005) for an argument why optimization methods are useful even if the genetic architecture was fully known.

challenges in the best possible way. In other words, individuals are assumed to behave optimally (given some set of constraints). In the current chapter, we will have a look at optimization models, to see what insight they can yield.

4.1 Why be honest?

Males of numerous species appear to try to 'impress' a female with elaborate displays. Why this might be happening has puzzled researchers ever since Darwin (1871), who lamented that seeing the peacock's fancy tail gave him a headache. The reason for the headache is that exaggerated ornaments do not seem to make much sense in the face of natural selection: they can only conceivably diminish a male's survival. Darwin's answer was to hypothesize that females possess aesthetic senses that favour ornamented males. The idea lay dormant for quite a while, and Darwin never tackled the question why females should have developed such preferences. After all, by mating with ornamented males they gain male offspring who inherit the damaging trait. Fisher (1930) provided great insight by pointing out that, firstly, the trait could very well be linked with viability and this can kick-start a female preference, and, secondly, even if female preferences subsequently drive the male trait to exaggerated levels that are no longer beneficial to viability, female preferences can be maintained in the population. This is because genes for preferences and genes for male traits end up in the same individuals, and the male trait can be selected for. The male trait in its exaggerated form is in the end not damaging to male fitness: precisely because there is a preference in the population, males with exaggerated traits enjoy much higher mating success than other males.

But does the preference sufficiently counteract viability costs? Although Fisher was very mathematically adept, he did not model what the net effect of these opposing forces are. Later, his insight was confirmed using very similar methods to those described in Ch. 3, though with an added layer of complexity: an explicit treatment of genetic covariances between two traits (female preferences and male traits; for a review, see Mead and Arnold (2004)).

Fisher's idea works, but modelling has led to a lot of insight (and some heated debate) about when exactly preferences can counteract costs on male viability (for a summary see Kokko *et al.* (2006a)). By now, models also consider additional factors that can influence female choice (reviewed, for example, by Andersson (1994), Kokko *et al.* (2003), Shuster and Wade (2003) and Andersson and Simmons (2006)). Great effort has been spent in trying to find out exactly what it is that females are after – or are they merely 'seduced' to mating maladaptively (Holland and Rice 1998; Gavrilets *et al.* 2001; Arnqvist and Rowe 2005), perhaps because males exploit the sensory responses of females (Arnqvist 2006). One intriguing finding that has emerged is that male traits are often condition-dependent; for example, better-fed dung beetles *Onthophagus taurus* court females more, and achieve higher mating success, than those that have not been fed (Kotiaho *et al.* 2001). Condition will, of course, depend on the environment in which a male was reared (e.g. the amount of food can vary), but it could also be related to a male's genetic 'quality'. Quality might refer to his immunocompetence in a world teeming with parasites, or perhaps his foraging efficiency, either of which could have a genetic component. It may be in a female's interest to mate with a high-condition male for two distinct reasons. Firstly, his good condition could be directly important to the female's breeding success, for example when the male provides parental care (male sticklebacks in poor condition sometimes eat all the eggs they are caring for (Candolin 2000)). Secondly, if condition is heritable, a female could be interested in finding good genes for the offspring (together with the associated attractiveness benefit of male offspring).

So, if traits correlate with a male's condition, the job is made easy for females: simply pick the best-looking trait, and voilà! one has found a mate who is in a good condition. But *why* do male traits obey condition dependence?

The question may sound silly. Of course males in better condition have the stamina to sing better (e.g. acoustic moths: Brandt and Greenfield 2004), drum more (drumming wolf spiders: Kotiaho *et al.* 2001) or develop eyes that are farther apart (which is 'sexy' in stalk-eyed flies, Cotton *et al.* (2004)). But is it obvious? The reason why condition-dependence deserves some formal analysis is that the allocation of one's resources to sexual signalling appears to be something that males can 'choose'. 'Choice' here has nothing to do with free will; it is most likely genetically controlled, analogous to the barnacles' responses to the environment described in Ch. 3. Consider a drumming wolf spider, which attracts females by rapidly beating his abdomen against dry leaves. This requires a lot of effort. A hungry spider will enter a physiological state that makes him drum less often (Kotiaho *et al.* 2001). 'Choosing' to drum less appears to be the appropriate way to deal with a trade-off between time spent drumming and time spent foraging, but we could also imagine alternatives such as 'drum at maximum speed until you drop dead'. Put more scientifically, males should possess *reaction norms* (Schlichting and Pigliucci 1998) that dictate how much of the available resources are spent in sexual ornamentation or displays. This is away from the male's own self-maintenance. The resources can be energy or whatever it is that is needed to develop the traits. The resource in question may vary from case to case, and in some cases specific physiological pathways can be involved. For example, there may be a limited supply of carotenoids, which are useful for developing immunocompetence – so maybe eating carrots really is good for you – but they are often used in sexual displays too, such as the red belly of male sticklebacks.

So, how exactly should a male of a particular condition allocate his resources? Let us be generous towards males and assume that they are quite free to allocate resources in whatever way they 'want'. In more precise words, we assume that their genetic architecture provides them with rules that find the optimal allocation for each condition. Optimal behaviour could in practice be implemented using very simple rules, for example 'when too hungry, stop drumming'. We do not assume decision mechanisms that are too sophisticated here, only that different rules can be produced by the available genetic architecture. To produce the 'drum until you drop dead' rule, for instance, it is conceivable that an animal could be hormonally or neurally wired to drum as long as there is some glucose and oxygen left for the muscles to operate at all. Whether that is optimal is another story, and that is what we would like to find out.

Now if females cannot detect condition directly but must rely on the male's display to infer condition, then we must ask what prevents a low-quality male (i.e. one in poor condition) from displaying so strongly that he looks equally good in the female's eyes as a better male. Why shouldn't a spider keep on drumming no matter how hungry he is? After all, without drumming he will not mate, and without mating it's 'game over' for him. But if all males keep drumming at an equal rate until they drop dead from exhaustion, they will all sound identical to females, and female preferences become pointless. Is there anything that prevents such 'cheating' strategies from spreading?

Let us consider this question using an *optimization approach*. Once again, we must remind ourselves of the inherent assumptions of a method. As the name suggests, optimization assumes that we can expect natural selection to find an optimal (fitness-maximizing) phenotype in each situation. This is not an obvious assumption: it requires that we believe that the genetics work flexibly enough that optimal phenotypes can be produced. In the current context, we should have faith that evolution can produce a physiological mechanism that has an introspective look at a male's own condition and consequently decides, 'given what the situation looks like, now let's try to drum every 4 minutes'. How likely this is in each case is a source of much debate. Often, it is an empirical question: we know that reaction norms do exist (Schlichting and Pigliucci 1998), and one can then study how finely tuned they are to respond to a variety of challenges. In general, we can expect evolution to produce more fine-tuned answers to frequently occurring variations in challenges, whereas optimal responses to novel or rarely occurring challenges are much less likely. For example, animals often misbehave grossly in environments altered by humans: mayflies use polarized light to detect pools to lay their eggs in, which in the current world can make them lay eggs on dry asphalt roads that also reflect polarized light (Kriska *et al.* 1998).

Often, we do not know exactly how 'optimal' a behaviour is expected to be. But the point of an optimality approach is often different. When we ask 'do we expect male displays to correlate with condition', digging into the genetics of reaction norms of this or that species is not the most urgent task. The question is more general than that. Do we expect sexual displays to correlate with condition, *assuming* that we give all males free hands and let them try to cheat if it benefits them? Odd as it may sound, giving hypothetical males such freedom is a *conservative* assumption when it comes to explaining why male traits should be honest signals of condition. We could, instead, investigate some forms of genetic or physiological constraints that simply prevent the poor-quality males from displaying strongly, but then we would not find out if the ornaments are still expected to be honestly correlated with condition if such constraints happen to be absent in some species. This is the fundamental reason why an optimization approach, giving the individuals completely free hands to decide what is best for them, is valuable in our case.

Optimization is all about trade-offs: maximizing fitness under constraints. The dilemma that a male faces is the following. If he invested all the resources (energy, carotenoids or whatever) he has in good looks (or persistent drumming), females would like him, but this is of little use if there's no energy left for self-maintenance, because he won't survive to enjoy his high mating success. Result: no fitness. If he invests all in self-maintenance, he will probably survive; but this is of little use in an evolutionary sense, because females will ignore him as he has no sexual traits. The best thing to do is clearly to invest some fraction of the available resources to self-maintenance, and some to developing the ornaments. But how much? And, crucially, how does the amount differ between males of different qualities (condition) – do they arrange themselves neatly and 'honestly' such that best males have best-looking traits? In other words, is it really *optimal* for a poor-quality male to have poorer-looking ornaments?

4.2 Simple cost–benefit analysis

Let's consider this question first with a simple model where the trade-off is depicted as a balance of costs and benefits. We want to think about the fitness of males who differ in condition and in the 'size' of their sexual trait. 'Size' will be interpreted broadly here; for example, a high drumming rate in a spider qualifies as a 'large' trait equally well as a long feather ornament in a bird. To provide the simplest possible answer to the question, 'should a male in good condition produce larger traits?', we can focus on two males: a high- and a low-quality one (i.e. in good or poor condition). The high-quality male will display a trait T_H, and the low-quality one will display the trait T_L. But we are not pre-assigning values to these traits. Rather, they can take any non-negative value, and we will then evaluate which choice of T yields the highest fitness for each male. Then, if we can derive the prediction that $T_H > T_L$, we have shown that honesty rules.

Why do we consider two males only, and not a whole continuum of possible conditions? This is simply to aid comparison. As we will see (p. 72), we can very easily vary the condition that one (or both) of these males have, thus considering a whole set of conditions and how that influences the trait.

Now, we need to specify the fitness of the males. This obeys a trade-off between benefits and costs, as described above. The benefit of having a large trait value is clear: females will mate with such males more often. This benefit could be, for example,

$$b(T) = T \tag{4.1}$$

The benefit (mating success) is here assumed to be proportional to the male trait. The notation with the bracketed T tells us that b is a function that depends on T. If this is clear from the context, it can also be left out; $b = T$.

Why do we assume a proportional increase, rather than any other increasing function, such as $\sqrt{T}$ or T^2? There is no good answer except that one has to begin with something. In general, there is a difference between models that aim to show that 'something is possible' (e.g. honesty can evolve, under some conditions), and those that aim to show that something is impossible (e.g., 'the evolution of honesty/altruism/ *insert your favourite topic here* can never happen in the way my colleague claims'). In the latter case, it is, of course, not sufficient to present one simple example where something does not happen and draw overly general conclusions. Much more general analysis is warranted, for example 'any increasing, non-negative function will lead to the same impossibility'. Typically, the mathematics that are needed to achieve such statements are somewhat more advanced – but not necessarily prohibitively so.

But back to our simplistic example. What the argument $b = T$ really does is to define the scale in which we think about the male trait. In effect, we have chosen to *measure* the size of the trait through its effects on mating success. Thus, an increase in the trait that leads to doubling the mating success is *defined* to be a "twice as large" trait. It is not trivial to give a scale to, say, the 'size' of bird song: male birds can differ in their repertoire size as well as in how much time the male spends singing, so our choice is to simplify everything to a scale that is based on how deeply females are impressed. But if this bothers you – empiricists will be quick to tell you that in this or that species song complexity varies yet has no effect at all on female choice – you can also decide that the traits are firmly measured as they were originally (in millimetres for the length of a feather ornament, or drums/minute for a spider). The relationship $b = T$ then states an assumption that females like male traits proportionally to their measured size. This immediately excludes species where the trait is unrelated to mating success. And, of course, you can change this assumption if you like.

Importantly, the benefit relationship $b = T$ is assumed to be the same for both high- and low-quality males. This reflects our assumption that females cannot detect male condition ('quality') directly; they can only rely on the trait as an indicator of condition that may or may not be honest. Therefore, a low-quality male will get equally as many females as a high-quality male, if both have equally large traits.

But when it comes to the costs, the males differ. We here think of the low-quality male as someone who suffers more if he tries to develop a certain size of trait, T. This means that the costs he pays are larger for each trait size. For example,

$$c_L(T) = \alpha_L T^2 \tag{4.2a}$$

$$c_H(T) = \alpha_H T^2 \tag{4.2b}$$

The biological justification for accelerating (here, squared) costs of trait production is that eventually, with ever increasing traits, the costs must exceed any mating benefit. There must be an ultimate limit of how fast and continuously a spider can drum, likewise peacocks cannot produce tails that are kilometres long. No male will enjoy positive fitness if he tries to do this; the effort will kill him long before this is achieved. The parameters α_L and α_H are coefficients that determine how large the costs will be for a given value of T. For example, when $T = 1$, the cost will be exactly α_L for the low-quality male and α_H for the high-quality male. Because the low-quality male is the one who suffers higher costs, we must have $\alpha_L > \alpha_H$.

What is fitness, then? The usual letter for fitness is W (although I have never met anyone who knows why). We now state that fitness equals benefits minus costs:

$$W_L(T) = b(T) - c_L(T) = T - \alpha_L T^2 \tag{4.3a}$$

$$W_H(T) = b(T) - c_H(T) = T - \alpha_H T^2 \tag{4.3b}$$

This specifies the relationship between trait size and fitness, for each type of male. If you frown and say 'why exactly benefits minus costs', read on; we will return to this question in Section 4.3. Assuming, for a moment, that we are happy with this conceptual choice, we are now ready to ask, how can we find the best choice of T for each of the males?

If you are analytically inclined and know already how to search for maximum values of functions, you may use those tools and jump to p. 72. We will join you there, once we have first had a look at what the maximization means, using some examples for values of α.

For example, let us consider $\alpha_L = 2$ and $\alpha_H = 1$. Again, these values are there just to start with; to reach more general conclusions, we will eventually have to do more. We can now plot the functions $b(T)$, $c_L(T)$, $c_H(T)$, $W_L(T)$ and $W_H(T)$, and see what they look like (Fig. 4.2). Here, we follow our assumption that each male will develop the trait up to the point where fitness is maximized: where the difference between costs and benefits is largest. (Note that this is *not* the point where the lines depicting

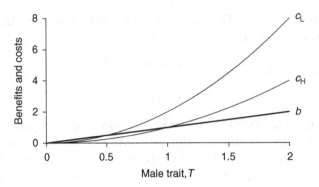

Fig. 4.2 Costs and benefits in the simple analysis; figure created as in Box 4.1.

b and *c* cross; we are instead looking for a position where *b* exceeds *c* and the gap between *b* and *c* is at its widest.)

Box 4.1

Finding the maximum difference between benefits and costs, here as a script. Running this script will produce Figs. 4.2 and 4.3.

```
T = linspace(0,2,101);
b = T;
cL = 2*T.^2; cH = T.^2;
figure(1); plot(T,b,T,cL,T,cH);
xlabel('Trait'); ylabel('Costs and benefits')
WL = b-cL; WH = b-cH;
figure(2); plot(T,WL,T,WH);
xlabel('Trait'); ylabel('Fitness');
grid % this draws helpful lines to see where the
% maxima are
axis([0  1  0  1]) % and we can also zoom in to have a
% closer look
```

But which values of T should we concentrate on? Our scale is quite abstract, so initially we have no clue. Trying out trait values between 0 and 2, just for a start, shows clearly (Fig. 4.2) that beyond $T = 2$ the costs exceed the benefits and increase much faster too, so there is no chance to find optima there (see Box 4.1 for the MATLAB code). In fact, Fig. 4.2 makes it clear that the benefits (thick line) exceed costs only where T falls

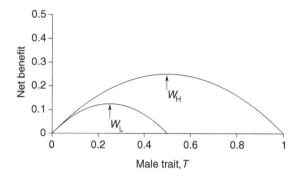

Fig. 4.3 Fitness as a net benefit, calculated for low-quality and high-quality males. Arrows indicate fitness maxima. Note that the arrow of the low-quality male occurs at a lower trait value than the arrow of the high-quality male.

below 1; hence in Fig. 4.3, which plots net fitness (benefits minus costs), we focus on this area.

Here, we immediately see that the low-quality male, who suffered higher costs when developing the trait, opts for smaller traits (Fig. 4.3). In other words, costs ensure honesty: even though we gave males the option to cheat, the costs of developing a dishonestly strong trait make this option one to be selected against. Low-quality males do, however, perform some levels of sexual displays: we did, after all, assume that a modest trait yields some (modest) mating success. This is quite reasonable: female preferences rarely operate in a one-off fashion that would only give matings to the very best males ever present in a population, for example because finding the male of one's dreams would entail too high travel costs for most females.

How to generalize this finding? We have only tried two values of α, which we simply pulled out of a hat without thinking much about the generality. Remember the salmon of Ch. 2? Again, it is very wise to try to establish an analytical solution. How does the optimal trait value depend on α, in general? So let us forget about these two particular males and write generally, $W(T) = T - \alpha T^2$. This is a differentiable function of T, which loosely speaking means that there are no abrupt changes or boundaries, just simple, nice, smoothly bending curves. ('To differentiate' means 'to take the derivative', which is impossible if there are abrupt changes in a function.) Because our function is differentiable, we can use elementary calculus, which dictates that local minima or maxima of a

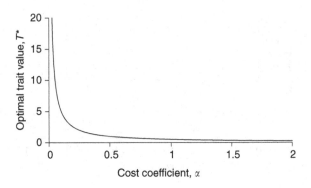

Fig. 4.4 The relationship between α and the optimal trait value T^*, as given by Eq. (4.5).

differentiable function can be found by looking for points where the derivative of a function is zero (Fig. 4.1).

The derivative in our male trait example is

$$dW/dT = 1 - 2\alpha T \tag{4.4}$$

Setting this to 0 yields a local minimum or maximum, which occurs at a value of T that we denote by T^*:

$$1 - 2\alpha T^* = 0$$

$$T^* = \frac{1}{2\alpha} \tag{4.5}$$

Instantly from this equation, we can see that the optimal trait size 0.25 follows for $\alpha_L = 2$, and 0.5 follows for $\alpha_H = 1$. We can also plot T^* against α and see that it is a decreasing function of α (Fig. 4.4).

If your life depended on getting this result right, you should still do a few more checks. Firstly, even though Fig. 4.3 shows quite convincingly that the condition is a local maximum rather than a minimum, it should be verified that this is the case for all values of α. Why? Because the condition $dW/dT = 0$ in general does not guarantee that fitness has peaked. A flat horizontal point along a curve can also occur at the bottom of a valley. The fact that, in our case, $dW/dT = 0$ finds a local maximum rather than a minimum is checked by taking the second-order derivative of fitness W. The idea of a second-order derivative (also called second derivative) is a bit complicated, but not too much. What we have simply called 'derivative' so far is the first-order derivative. This is the rate of change of a function (i.e. its slope: positive derivative means an increasing function). The second-order derivative is the derivative of the derivative, in other words, how quickly does the rate of change change. Sounds repetitive? Changes in the rate of change give information about accelerations and decelerations of functions, which is why second-order derivatives are useful. For example, a positive first-order derivative of a function, combined with a positive second-order derivative, means that the function is growing and growth is also speeding up as we go (exponential population growth is an example). Other combinations are possible too, and let's have a look at what happens at our $dW/dT = 0$: the first-order derivative is zero, and the second-order one is obtained by taking the derivative of $1 - 2\alpha T$ with respect to T, which yields the result -2α. This is negative for all positive values of α (which is what we are considering here).

So, when the first-order derivative is zero and the second-order derivative is negative, it means that the rate of increase of fitness is not only zero but also declining. Something that declines from zero must become negative. So, fitness 'grows negatively', declines, if we consider values of T to the right of the point T^* (at which $dT/dW = 0$). Also, if the rate of fitness increase is declining when it is zero, it means that just before zero it must have been positive. So, considering values of T to the left of T^*, fitness must have been growing (positive rate of increase). This is the telltale sign of a maximum: from the left, an increase, to the right a decrease, exactly as in Fig. 4.3. It all can sound tedious, but it can be summed up concisely: $dT/dW = 0$ and $d^2T/dW^2 < 0$ (that's the notation for second-order derivatives) implies that we are staring at a local maximum.[3]

Secondly, we should remember that setting a derivative to zero does not find maxima that occur at boundaries beyond which we do not wish

---

[3] If you are a person who finds it easy to think in terms of physical analogies, imagine throwing a ball vertically upwards and denote the height of the ball at time t with a function $h(t)$. The first-order derivative of the distance, dh/dt, is the ball's vertical speed. This is positive if the ball still flies upwards, negative if it is already returning back. The second-order derivative d^2h/dt^2 gives the acceleration of the ball. This is constant and negative all the time, since gravity has a negative effect on a speed that is defined to be positive when going upwards. Negative acceleration while the ball is moving upwards means that the ball decelerates. The same negative acceleration while moving down means that the distance between the ball and the ground diminishes ever faster. This is indeed how balls behave, and the maximum height is reached when the negative acceleration has 'eaten up' all the initial speed that the ball had. At this point $dh/dt = 0$ combines with $d^2h/dt^2 < 0$, so the height reached is a maximum. Physics textbooks have very similar graphs as our Fig. 4.3, but the x axis denotes time, and the y axis gives the height the ball has reached.

to look. We have one such boundary, the "no trait" case $T = 0$: no spider can drum less than zero times per minute, so we wish to stop the scale there. Although not the case in Fig. 4.3, it could in principle happen that the best thing that a spider can do is to stop drumming altogether (e.g. during a time when no females are around, but predators are listening). In such a case, fitness is highest at $W(0)$ and declines ever after: the less drumming, the better, and $W(0)$ is a maximum even though it is not located at a 'flat' part of the curve where dW/dT would equal 0. To avoid overlooking cases where $W(0)$ might be the best thing to do, and knowing that $dW/dT = 0$ will only find maxima that are 'flat' (this is how it finds us the value T^*), we should consider both $W(T^*)$ and $W(0)$ as candidate maxima. We can then decide which one is higher. Now $W(T^*) = 1/(2\alpha) - \alpha[1/(2\alpha)]^2 = 1/(4\alpha)$, which is greater than $W(0) = 0$ for all sensible values $\alpha > 0$. So, since in our case we are assuming that some females are paying attention, $W(T^*)$ always outperforms $W(0)$, but it was good to check it to make sure.

Finally, you should take the derivative $dT^*/d\alpha$ and verify that it is always negative (for sensible values $\alpha > 0$). Why? This ensures that T^* really always, always decreases with α so that the guys with big traits are the ones who pay the lower costs. So we can be quite confident about our conclusions...or can we?

4.3 A more comprehensive model based on life history

If you feel slightly annoyed by the model of Section 4.2, that is for a good reason. It is certainly an oversimplified model. Simplicity as such can be a good thing, but not if the biological interpretation becomes shaky as a result. For example, if $\alpha = 1.2$, and the male possesses a trait $T = 1.2$, it has to pay a cost $c(T) = 1.728$. What is the biological meaning of this abstract number? Worse still, the net fitness of the male is negative, $W = -0.528$, and it is not exactly clear what this should mean. One could, of course, say that zero fitness should be interpreted as a kind of a baseline value, and negative values mean that the male won't have much chance to survive at all under these circumstances. But this is handwaving and feels unpleasant for a good reason. As we shall see here, too much reliance on simple 'benefit minus cost' arguments can be misleading, and people indeed have been misled in the past. This is why it is important for empiricists too to learn enough modelling: one needs to have enough understanding of what a model subject has eaten to be able to assess what the assumptions and predictions truly mean. And, if necessary, one should also have the courage to ask whether the life history of an animal has been appropriately modelled, while at the same time accepting that trying to include every detail is pointless.

Let's take improved realism as a goal, by quantifying the survival chances more directly and developing a proper measure of fitness. There is an additional reason why more modelling effort is needed, even after the above models illuminate a basic prerequisite for honesty (sexual signalling should impose differential costs such that poorer quality suffer more from increasing traits). The reason for more work is that we hardly ever know what shapes the trade-offs really take in nature; sometimes detecting that there *is* a trade-off can take years of effort. It is, therefore, good to check what happens if the allocation of resources happens in a slightly different way than the abstract and oversimplistic form of Section 4.2.

So, let's make different assumptions and be explicit about how well a male survives. This is done by expressing the cost of a sexual trait as a reduction in condition, which is then reflected in annual survival. Now we imagine that males differ in the total amount of resources available to them (R). Males with high R are high-quality males, and our question is, do they show this by signalling more intensely (larger male trait T). We can now assume that the trait T is measured in units of how much resources its development has required. This means that T cannot exceed the amount of resources that a male has (R).

Survival is then something that increases with R but, for any given R, it decreases with T. A neat way to model this trade-off is to introduce the idea of net condition (i.e. condition after sexual signalling has taken its toll). The male's net condition C equals the amount the male spares to himself in the allocation to sexual displays versus self-maintenance: $C = R - T$. Survival S is an increasing function of net condition C. Since $0 \leq T \leq R$, it follows that net condition C will lie in the range $0 \leq C \leq R$. An identical way to express the trade-off is that a male with resources R will choose a fraction aR to become his trait, and a fraction $(1-a)R$ is left to become his net condition. We must have $0 \leq a \leq 1$, for this to make sense: eating cakes and having them too is not allowed here (Fig. 4.5). A male is assumed to be able to change the allocation a but not the total resources R that is his lot in life.

How should survival S depend on condition, C: in other words, what should the function $S(C)$ look like? Of course, once again we do not know what exact shape this takes for any organism, but several points can be made based on biological realism. A sensible survival function should,

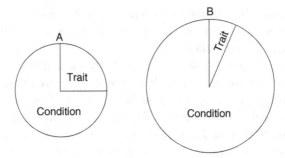

Fig. 4.5 Hypothetical resource allocation in two males. Male A has a small amount of resources to spend (the total area of the 'cake' corresponds to R) but allocates a fairly large fraction $a = 1/4$ of resources to developing the sexual trait. Male B has more resources (a larger cake), but the allocation fraction is smaller. The area of the slice aR corresponds to the trait, and the remaining area indicates the male's condition (C). Since large values of C improve male survival, we can predict that male A finds more females but survives poorly compared to male B.

first of all, never exceed unity, since it is not possible to survive with, say, 120% probability. It, therefore, makes sense that the function has a saturating shape. Also, it is conceivable that very low condition means very slim survival chances. Therefore, we start imagining an S-shaped (sigmoidal) function that could capture our ideas well. A suitable shape is given in Fig. 4.6, described by

$$S(C) = \frac{1}{1 + \exp(-10(C - 0.5))} \tag{4.6}$$

It is a function with many nice features: it is a shape that resembles the growth of a population experiencing density dependence. In fact the shape is identical to the predicted growth of a population in which the *per capita* growth rate (i.e. births minus deaths calculated on an individual basis) falls linearly with density, and growth is continuous, as opposed to discrete breeding seasons. This is called logistic growth in continuous time.[4]

---

[4] The 'discovery' of the logistic growth function is usually attributed to Raymond Pearl, whose career was hit by a devastating fire in 1919 that destroyed his mouse-breeding unit and his collection of literature, unpublished records and data that had taken 20 years to put together. He consequently turned his attention from tuberculosis and public health to the abstract patterns of population growth – so there are many routes to becoming a successful theoretician! Like many good anecdotes, however, this one too appears to be an oversimplification: some empirical *Drosophila* breeding happened too between the fire and the logistic function, and the function itself had made brief appearances in the literature before Pearl's work. See Kingsland (1995) for details.

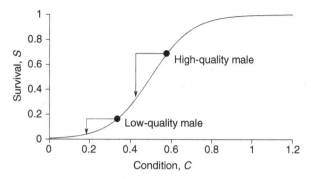

Fig. 4.6 A sigmoidal relationship between male condition C and survival S, here exemplified with $S = 1/(1 + \exp(-10(C - 0.5)))$. If males did not develop sexual displays, they would survive according to the location of the filled dots, where $C = R$ (examples given for two males differing in resources available, R). The size of the sexual trait is given by the length of the leftwards arrows, and the consequent cost of reduced survival is the downwards arrow. Achieving a display of the same magnitude can lead to a larger reduction in survival for the high-quality than for the low-quality male.

To get from the linear decline in the *per capita* growth rate to the S-shaped overall pattern of population growth is a frequently repeated exercise in ecology textbooks. For example, Case (2000) gave a derivation with the notation $N(t) = K/[1 + (K-N_0)/N_0 \exp(-rt)]$, where $N(t)$ is the population size at time t, K the carrying capacity and N_0 the initial population size. It may not be immediately obvious that $S(C)$ and $N(t)$ have the same S-shape, but they do: $S(C)$ can also be written as $1/(1 + \exp(5 - 10C)) = 1/(1 + \exp(5)\exp(-10C))$, which makes the equivalence clear when $K = 1$, $(K-N_0)/N_0 = \exp(5)$ and $r = 10$.

The assumptions of the model can now be summarized in one figure (Fig. 4.6): there are males who differ in their resources R, and they are, therefore, initially located at different points of the survival curve. They can choose to spend any amount of resources in sexual displays T (but only up to $T = R$), and this will reduce their net condition and hence cause a drop in annual survival. Sexual displays are beneficial since they impress females, and the annual number of matings (which is assumed proportional to offspring production) is M, which again is proportional to T... why? For no better reason than that it may be best not to change every assumption of the previous model at once in order to be able to pinpoint the cause of any changes in the results.

So ... how impressive should the traits look or sound like?

In principle, we proceed exactly as we did before in Section 4.2. We write down an expression for fitness and look for the maximum. But, hang on, what *is* fitness? We certainly want to avoid the oversimplistic 'benefits minus costs' approach, but illogical quick fixes such as 'survival + mating success' are rejected as quickly as they come to mind (think about how many offspring a male will have if he survives well but never achieves any mating success). Instead, we must pause for a moment to think how many offspring the male is about to get in his lifetime.

At this point, let us stop using the drumming spider example as our sole guiding light. To reach more general conclusions, we should probably give our males the chance to live for more than 1 year – while also allowing them to squander it all in one super-sexy mating season if they so wish. If the male gains M offspring (or a number proportional to M, and hence proportional to T) every year, his lifetime offspring production must be proportional to M times the expected number of years he can participate in mating activities (i.e. be alive). Therefore, we are keen to know his expected lifespan. We must also consider whether we think of pre- or postreproductive costs of sexual advertising (cf. Jönsson *et al.* 1998). In the case of prereproductive costs, survival falls in the juvenile life stage already if the male is about to use a lot of his resources in sexual signalling. But if costs are paid afterwards, juvenile survival will not be influenced. There is no 'correct' choice here, and instead we must *decide* which case we wish to consider now. Perhaps it is more commonly the case that costs are only paid once the display has taken its toll on the male's condition, so let us focus on the latter type of costs. This means that we can ignore variations in juvenile survival and consider a male's fitness from maturation onwards: a male that is just about to enter his first mating season. As always, if you would have liked to make different choices ... feel free to alter the equations accordingly, and look what happens.

Our newly matured male will first gain the mating success from one mating season (M). Then the male's annual survival from the end of one mating season to the start of the next equals S. If he survived, he gets M again, and so on. Note that we have chosen to assume that gaining the first M is guaranteed; therefore, we are implicitly assuming that no significant mortality occurs *during* the mating season. This brings us further away from the spider example: high mate-searching activity in wolf spiders directly increases the risk that the male is predated (Kotiaho *et al.* 1998), so looking at such types of mortality would necessitate an alternative model. Now we also notice that we have made the implicit assumption that the male invests equally much in the trait every year of his life: we have not thought at all about age-dependent variations in S, R, T, M or C. For simplicity, let's keep it that way and assume that the male survives with probability S every single year of his life. So, the task is to derive the expected lifespan of an animal from maturation onwards when annual survival probability equals S each year after maturation.

Some maths is now needed. As said, the male is guaranteed to experience his first mating season because variation in juvenile survival was irrelevant and we can concentrate on males who already have survived until maturation. The probability that the male dies before he ever has a chance to experience *another* mating season is $(1 - S)$; this is a failure to survive his first year post-mating (survival probability S). The probability that he experiences exactly one additional mating season is $S(1-S)$: this means surviving (probability S) the first year, but dying (probability $1-S$) thereafter. The probability for experiencing exactly two additional mating seasons is $S^2(1-S)$, three additional ones, $S^3(1-S)$, and so on. The expected lifespan after the first mating season, then, is a sum of an infinite number of terms,

$$(1 - S) \cdot 0 + (1 - S)S \cdot 1 + S^2(1 - S) \cdot 2 + S^3(1 - S) \cdot 3 + \cdots$$

which is a weighted sum of all possible post-maturation lifespans (0, 1, 2, 3, ...), weights being the probabilities of having exactly this lifespan. This can be expressed more concisely as

$$\sum_{i=0}^{\infty}(1 - S)S^i i \tag{4.7}$$

which means 'add together $(1-S)S^i i$ for all values of i from zero to infinity'. Looks nasty? Very luckily, sums of an infinite series quite often *converge*, which means that despite the infinite number of terms the result is a single, well-defined value. In this case, it is $S/(1-S)$. Wow, how did we get that? It is an example of what is called, erm, that slippery concept of mathematical beauty. These kinds of result have excited mathematicians for millennia: something complicated turns out to be equivalent to something much simpler. A mathematician is always very pleased when he or she can prove something like this. Ever afterwards, we can take the result for granted, without need to go through the tedious calculation. If it worries you that you might not have found this result had it not been given here as if it had just fallen from the sky, take solace in a couple of reassuring facts. One is that programs such as MATHEMATICA and MAPLE know most of the necessary rules, so you can ask them. The other one is that mathematical modelling in ecology typically requires very, very few infinite sums. Most modellers will ever only need this particular one – if any. Therefore, the most useful rules keep appearing in the literature so often that one becomes quickly familiar with them.

So, the expected lifespan is $S/(1-S)$. You may wish to verify that this is an increasing function of annual survival, S: and it has an interesting shape too. Given that a male's mating success is M, which, remember, is proportional to T, the male's lifetime fitness is proportional to

$$W(T,R) = M(T) + M(T)\frac{S(R-T)}{1 - S(R-T)} \tag{4.8}$$

Where did this come from? The first M refers to mating success in the current season that our newly matured male is about to start. The second term is his expected future reproductive success: every year he should gain M matings, and the expected number of future years is $S/(1-S)$. The expression in parentheses $(R-T)$ 'input' (argument) of functions: remember that the value of S depends on condition C, and it must be evaluated assuming that C equals $R-T$. The expression in parentheses (T,R) likewise reminds us that fitness depends on the values of T, which the male can choose, and R, which the male cannot choose (but it can vary among males). Mating success depends on T only, because females cannot detect R directly.

A remark. Do not let the language of proportions confuse you: this is simply to get rid of some extra parameters that will not influence the outcomes anyway. For example, we could include a proportionality constant such that $M = \beta T$. But an examination of Eq. (4.8) reveals that this would simply multiply all the values of W by β, in other words, change the scale of the y axis of any graph where we plot W against trait values T. Such a change of y axis scales cannot alter the *horizontal* (x axis) position of the best value of the trait, T^*. It will only amplify the vertical differences that we are not really interested in – so we have good reasons to ignore such proportionality constants.[5]

---

[5] Had we started off thinking that β is important and, therefore, included it in the model, it would be useful to know about techniques that help to realize that it is superfluous. In our case, it would be quite easy to spot the fact that β has no effect on T^* at all, but for more general techniques see Ch. 9 in Otto and Day (2006).

Where is the best value of the trait T^* for each male? If we have settled on a particular function of $S(C)$, the task is to calculate W for each possible value of T, given a particular value of R. The possible values for T range between 0 and R. Once the best value (one that leads to highest $W(T,R)$) has been found, this is recorded as the best choice for a male with resources R. For sigmoidal shapes of $S(C)$, one cannot unfortunately find neat analytical solutions such as Eq. (4.5), but a numerical procedure does the trick dutifully (see Box 4.2 for the implementation using MATLAB). Note that if all males have incredibly large quantities of the resource, they can invest so much in their survival that they become real Methuselahs – so it is best to check what their expected lifespan is; hence we calculate the variable Seasons in the program.

Box 4.2

The life history approach to the display problem. The program should be saved in a file called maledisplay.m before running it. It takes resourcerange as its input. This is a vector with two values: the lowest and the highest value of resources that we consider feasible in the population. Therefore if we run the program with resourcerange = [0, 0.2], i.e. by typing maledisplay ([0, 0.2]) into MATLAB, the output looks like that in Fig. 4.7.

```
function [Trait,Seasons,Fit,R] =maledisplay         } one
   (resourcerange)                                   } line
% function [Trait,Seasons,Fit,R] =maledisplay
% (resourcerange)
% 'resourcerange' should be a vector with two numbers
% - one for the low end of the range considered, the
% other the high end
% There are four outputs
% Trait   - the male trait (for each value of R)
% Seasons - how many mating seasons each male is
% expected to live
% Fit - the males' fitness
% R    - the resource values considered here; these
% will be 100 values between resourcerange(1) and
% resourcerange(2)
R =linspace(resourcerange(1),resourcerange(2),        } one
   100);                                              } line
```

Box 4.2 cont.

```
% the above lets us consider 100 different males,
% resources varying between the lowest and highest
% value given as inputs
for i = 1:length(R)
  % we consider the ith male, i.e. ith possible
  % resource value
  % possible trait values range between 0 and R(i)
  trait = linspace(0,R(i),1000);
  netcondition = R(i)-trait;

  % this is the function used in Fig. (4.6)
  survival = 1./(1+exp(-10*(netcondition-0.5)));

  % expected number of future mating seasons is the
  % result of the infinite sum given in Eq. 4.7
  futurematingseasons = survival./(1-survival);

  % mating success is just proportional to the
  % trait ...
  m = trait;

  % fitness now from eq. 4.8
  fitness = m+m.*futurematingseasons;

  % Matlab's 'max' function finds the maximum value
  % of a vector of numbers, and the location
  [bestfitness,bestindex] = max(fitness);

  % then we need to check which trait value produced
  % this best value ...
  Trait(i) = trait(bestindex);

  % ... and let's also calculate how long the male
  % survives in this case
  Seasons(i) = 1+futurematingseasons(bestindex);

  % ... as well as save the information on this male's
  % fitness when displaying optimally
  Fit(i) = fitness(bestindex);
end;
figure(1);
subplot(3,1,1);
```

```
plot(R,Trait,'g');
xlabel('Resources')
ylabel('Male trait');
subplot(3,1,2);
plot(R,Seasons,'b');
xlabel('Resources')
ylabel('Expected number of mating seasons');
subplot(3,1,3);
plot(R,Fit,'r');
xlabel('Resources')
ylabel('Fitness');
```

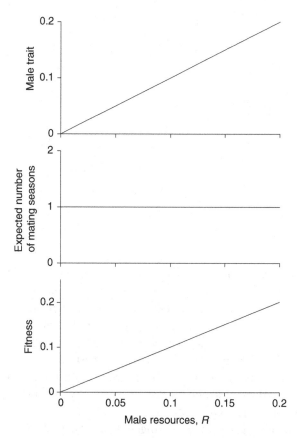

Fig. 4.7 Optimal male traits, the expected number of mating seasons, and fitness $W(T, R)$, for a range of R values between 0 and 0.2.

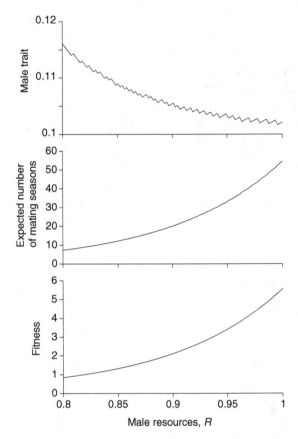

Fig. 4.8 Optimal male traits as in Fig. 4.7, but with *R* varying between 0.8 and 1.

Trying out a range of resources, *R*, between 0 and 0.2 leads to optimal solutions as given in Fig. 4.7. The interpretation is quite clear: male traits are exactly equal to the resources available (i.e. they have chosen to squander it all in one go), and the expected number of mating seasons equals 1 for any type of male. This is an example of a life history called *semelparity*: all males invest everything in their first ever mating season so that they simply die afterwards. The system is still honest: males with better resources display stronger signals, when everybody uses up all their condition in one single go.

A different choice of resource ranges, from 0.8 to 1, gives a totally different answer (Fig. 4.8). Shockingly enough, now the better-off males display *less* than their poorer comrades: increasing *R* corresponds to decreasing T. Males with high R nevertheless enjoy higher fitness, which cannot be explained by their choice of T but is due to a very, very strong increase of lifespan with R: more than 50 mating seasons for males with a whole one unit of resources.

There is some 'wobbling' visible in the male trait, but this is a matter of a limited accuracy with which solutions were computed: each male was given a slightly different set of values to choose from, as the 1000 choices between 0 and R vary between males with unequal total resources R. Some level of numerical inaccuracy cannot be avoided in numerical solutions, and in each case the question is whether they cause serious enough trouble to be worried about. They certainly do, if they accumulate in long chains of calculations where each result is based on the previous one – but this is not the case here. In our case, the wobble practically disappears if each male is given 10 000 different choices. The computing time obviously increases somewhat.[6]

The much more worrying aspect of Fig. 4.8 is the biological interpretation. What on earth is going on here? Nothing less grave than a proof that honesty is not always maintained. To gain an overall picture, let's depict all solutions between 0 and 1 (Fig. 4.9). Now we see that male traits increase with resources, until resources reach a threshold value around 0.8. Thereafter, a male can survive such a long time by displaying prudently; the optimal action is to reap the rewards slowly by displaying modestly in any given year. His eventual fitness is nevertheless higher than that of males with poorer resources. This makes sense because more resources should always yield higher fitness – it would indeed be silly if a male performed less well when given more chances to improve his fitness components.

Is honesty maintained overall? It depends. If there are no males whose resource levels exceed 0.8, the signalling system is honest. But if there are some males with resource levels near 0.7, and others near 1, then we may have the strongest signals in males that are near the low end of the quality spectrum present. Does this mean that our earlier insights (Section 4.2), that differential costs ensure honesty, are wrong? Not really; models are simply mathematical consequences of certain assumptions, and as such they cannot be 'wrong'. Instead, if there is a difference between the outcomes of two models, there must be something

---

[6] There are also much more sophisticated ways to find maxima or minima than the 'brute force' approach adopted here. Mathematical software has been developed with this issue in mind, and in MATLAB, one starting point is the function `fmin`. MATLAB even has a whole toolbox devoted to optimization algorithms.

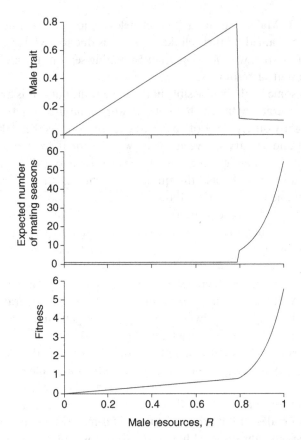

Fig. 4.9 Optimal male traits as in Fig. 4.7, but with R varying between 0 and 1.

different about the assumptions. Indeed, in Section 4.2 we assumed that low-quality males always pay higher costs when trying to increase their level of signalling. But with the current model, a careful look at Fig. 4.6 reveals that this assumption is not always valid. A sigmoidal survival function predicts that a high-quality male can have much *more* survival to lose by displaying than a low-quality male who survives poorly anyway. Thus, the latter model has taught us an important lesson: it may well remain true that steeper costs for low-quality males promote honesty, but who would have thought that costs, when given a proper life history definition, are not always larger for low-quality males.

There is, indeed, some empirical evidence that male display traits can be very flexible. If male sticklebacks find themselves in very poor condition, they may, counterintuitively, increase their red coloration to surpass any other male in the population (Candolin 1999, 2000). Conversely, high-quality male field crickets invest so heavily in sexual display that they die younger than low-quality males (Hunt *et al.* 2004). How is honesty then maintained, in general? If most males found themselves in the range of resource values between 0.8 and 1 in our example, it is expected that females stop paying attention to the signal that has become dishonest; and there goes the selective pressure that makes males display too. A signalling system cannot remain stable in this case, and there is a true threat that honesty can collapse and the whole setting of sexual selection disappears. But if only very few males enjoy such great resources that it pays for them to sacrifice some mating success for almost guaranteed survival from year to year (say, resource values range from 0.3 to 0.9, with only a few males exceeding 0.8), then the system is still 'honest on average'. This meaning that females mate with males with better resources if they rely on the display as an indicator of male resources than if they mate randomly.

This statement, by the way, is somewhat different to a claim I made some years ago (Kokko 1997), where I presented results that were very similar to Figs. 4.6–4.9, though in a context where signalling effort could vary over the lifetime of a male. I concluded then that, since males with best resources always end up having highest fitness, the presence of some cheaters cannot destroy the fact that the system remains always 'honesty on average'. This is an erroneous conclusion. Honesty, from the females' point of view, requires that display traits correlate with resources (or quality), not that the fitness of males correlates with resources. Funnily enough, nobody seems to have spotted this mistaken interpretation until now. The lesson: never stop being alert when reading published modelling papers.

To save my face, I might add that others too have overlooked important messages in models of honesty. Empirical attempts to test if conditions for honesty hold look for evidence that signals are costly to produce. A focus on the mere existence of a cost is a drastic over-simplification: even our simplistic cost–benefit analysis is sufficient to show this. Imagine that in Eq. (4.3) we add a constant K to the cost suffered by one of the males (say, the low-quality one). Positive or negative K would shift the fitness of this male downwards or upwards but would leave the horizontal position of the maximum unchanged. The simplified cost–benefit analysis (Section 4.2), therefore, would show that it is differential rather than absolute costs we should be looking for.

In other words, it is not sufficient to measure how much males suffer currently, we need to know how much males who differ in quality suffer if they attempt to *increase* their display. In modelling jargon, this is called a *marginal* cost. Empirical tests specifically addressing this issue are not that common, although, to be fair, many empiricists use expressions such as 'low-quality males are less able to bear the costs', which implies that the issue has been correctly understood.

But it gets worse. As said above, models cannot be 'wrong' because they only draw logical conclusions from their assumptions, but their assumptions can be irrelevant to real life. In particular, they can be oversimplified to the point that the conclusions become misleading. Thomas Getty (2006) has argued that the additive model of costs and benefits falls into this category, even when the results are phrased in terms of differential costs. Fitness is typically a multiplicative creature rather than an additive one. In this section, we managed to remember this by multiplying mating success with the number of seasons, and we found that honesty is a more slippery concept than we thought while still working our way through Section 4.2. Getty (2006) tackled the problem of multiplicative fitness by developing general arguments based on arbitrarily shaped functions rather than assuming on some predefined shape (see pp. 128–130 if such a skill appears mysterious). He showed that it makes more sense to speak about higher quality signallers being more efficient at converting advertising into fitness than to focus on differential costs only. In other words, there is no shortcut: one should create predictions on honesty by finding out how fitness as a whole varies with respect to the signal. This is indeed the approach that we have followed in this section. Conclusion: never stop asking whether the assumptions truly capture the essence of the biological problem. This is certainly a challenge, because one also has to avoid the opposite pitfall of clouding the issue with unnecessary detail, which could again make a model too limited in its applicability.

4.4 Got interested?

Optimization techniques have been used countless times when making arguments about adaptation (Parker and Maynard Smith 1990). Some of these applications consider a more direct proxy of fitness than others: a summary of the prospects of survival of parent plus offspring when analyzing optimal clutch sizes (Rowe *et al.* 1994) is clearly much more closely related to fitness than, say, minimizing the time it takes to complete a migratory journey (Weber and Houston 1997). It is consequently important to assess whether the *currency* that is being maximized or minimized is the correct one. Often we can argue that, *all other things being equal*, more food, or less time spent during migration, is a good thing. But are all other things equal? Obviously, good biological insight is required to make sensible assumptions regarding trade-offs. The more abstract the currency, the less clear it is what the biological meaning is (Sections 4.2 and 4.3). It is consequently easier to be convinced when the currency is a total count of lifetime offspring production than when it is simply thought to correlate with fitness but is clearly only one component of it, such as energy intake. However, counting offspring can lead to problems too, since there are many ways to count once one enters the realm of life history theory (see Brommer 2000; Brommer *et al.* 2002, 2004). A fully self-consistent model should, therefore, specify the full life cycle in its ecological setting, and we will return to this issue in Ch. 7. If this is done successfully, the currency becomes identical to the reproductive value of the organism, allowing very powerful inferences (McNamara and Houston 1986; Mangel *et al.* 1994; Houston and McNamara 1999). McNamara *et al.* (2001) have provided a nice and concise overview of many central questions for which optimization approaches (including methods that will be introduced in the next few chapters) have been used.

Optimization techniques find well-performing phenotypes. Phenotypic approaches (including an important special case, game theory, which will be dealt with in Chs. 5 and 6) are often placed in a contrasting position with genetic approaches, despite the underlying similarities described in the beginning of this chapter. The idea that evolution optimizes anything is hotly contested by some authors (for a classic reference see Gould and Lewontin (1979)). The main reason behind the criticism is that the ease with which adaptive explanations can be created can, in the worst case, lead to 'just so stories', by which one means an explanation that seeks adaptive value regardless of the likelihood that evolution can be driven by this benefit: a worst-case example is 'we evolved a nose to give support to spectacles (eyeglasses)'.

The crucial issue is to be able to make sensible choices: what can we expect natural selection to achieve, and what is too good to be true? For reasons that I will explain in Ch. 9 in more detail, I regard this largely as an empirical question. In this context, it is worth pointing out that the point of optimization modelling is not to prove that individuals of species

X can perform behaviour Y but to examine what sort of behaviours *would* be beneficial *assuming* the genetic machinery producing the optimal phenotype can be found (and that the selective advantage is strong enough to be of significance). The heuristic value of such an exercise should not be underestimated (Williams 1985; Rosales 2005), although some authors beg to differ (Sarkar 2005). Clearly, the task is to evaluate the evidence critically for and against the proposed adaptive machinery. For further discussion on the subject, see Parker and Maynard Smith (1990), Godfray and Parker (1991), Arnold (1992), Moore and Boake (1994), Weissing (1996) and Blows and Hoffman (2005).

5

Dynamic optimization

where we travel north,
and learn how to survive the winter.

The optimization procedures in the previous chapter were *static*: this means that decisions did not change over time, and the sexual displays of males looked the same mating season after season to the day they died.

Often, however, decisions are *dynamic*, which means that they change over time. For example, old individuals may face different trade-offs from young ones, because they find themselves in a different *state*: for example, they may be more experienced foragers and can, therefore, budget their time differently from young individuals. Or, environments may change over time: whether an insect or a copepod should develop directly into an adult, or first enter diapause, depends on whether summer is coming to a close. Very often, the optimal act is not one that maximizes immediate rewards – as any student doing long and difficult coursework can confirm. A copepod, for example, might gain some reproductive success sooner if it develops quickly, but if the reproductive season really is about to end, larger rewards await an individual who ignores this immediate gain and instead develops slowly, to mature at the beginning of the following season.

Dynamic optimization is needed when individuals have to make multiple decisions over time, and past decisions influence the options available in the future. For example, a student who has not yet spent enough time learning for the examinations that loom in the near future cannot expect to go to the movies with friends and also expect a good grade, whereas this option might be available to her friend who has started learning earlier. For our poor student, what is the best way to act from now on? A general principle of dynamic optimization (or dynamic programming,

91

as it is also called) is that *the best actions from now on can depend on the current state, but they do not depend on how one entered the current state.* In other words, it is true that a student who has not yet opened the textbook has to work harder just prior to the exam. But her performance is not improved if she spends time regretting past actions ('why did I not open that textbook earlier ... ') or concentrates on excuses for being where she is now ('it's unfair, my student loan ran out so I had to work at this fast food chain'). Regardless of why she has ended up in a state where the textbook material sounds like Greek, the best option from now until the exam date is to work hard – and go to the movies afterwards.

The principle of dynamic optimization may be good advice for life in general,[1] but it also makes calculations of complex pathways much easier. Consider the following real-life example. A student is about to start a PhD in a foreign country and has to travel from, say, Barcelona (the capital of Catalonia in Spain), to Jyväskylä in central Finland. He is keen to see a bit of Europe on the way, and there are many cities he would like to see – but, of course, visiting all of them is not really feasible. So he consults his friends, and together they graph several different alternative routes, which form a simple network (Fig. 5.1).

How to choose? There are many different trade-offs that could influence the student's decision. For example, our student has a friend in Paris who will be driving to Hamburg. They could do the trip together, which would be extremely good fun. Our student decides to be very methodological and scores expected 'happiness points' for each part of the journey. For example, the drive with the Parisian is worth 15 points. (Fig. 5.2).

Starting from the top end, the trips from either Turku, Helsinki or St Petersburg to the final destination do not score much: this is all mostly within Finland where our student will spend the next 3 years anyway, at this point during the journey he imagines he'll be happiest if he can make this final leg without much hassle. Since crossing the Russian

---

[1] Failure to adhere to this principle can lead to the *Concorde fallacy*. This is fallacious reasoning that follows the logic 'I cannot stop now, because otherwise what I've invested so far will be lost'. This makes no sense, if carrying on only leads to bigger losses than stopping, coming to terms with the past failure and moving on to something new. The name refers to the fact that French and British governments carried on losing taxpayer's money on funding the supersonic transport jet long after it was clear it would never be a commercial success (see Arkes and Ayton (1999) for further discussion). Note that some verbal explanations of sex roles and parental care seem to fall into this trap, when a female is argued to benefit from investing in current offspring more than the male because she has already invested more in them so far. The matter is not that simple (Curio 1987). Queller (1997) showed how to argue more logically about this issue (see also Kokko and Jennions 2003; Houston *et al.* 2005).

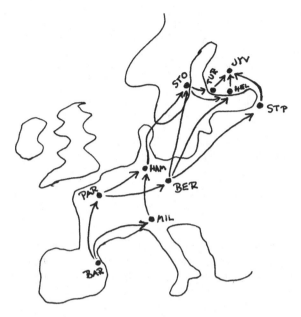

Fig. 5.1 A map of Europe as drawn on a napkin by a student trying to decide which route to take from Barcelona to Jyväskylä. Cities that might or might not fall along the way are Barcelona (BAR), Paris (PAR), Milan (MIL), Hamburg (HAM), Berlin (BER), Stockholm (STO), St Petersburg (STP), Turku (TUR), Helsinki (HEL) and Jyväskylä (JYV).

border involves some hassle, this leg gets '− 1' happiness point, and the other ones 0.

Arriving in each of these cities (Turku, Helsinki or St Petersburg) is a different matter. Seeing St Petersburg must be cool: 5 points. Also, arriving in Turku would be nice, as our student has heard that the ferry route from Stockholm would take him through a beautiful archipelago: 2 points, much better than the more boring Stockholm–Helsinki route (which still could be offering some fun, 1 point). And so on. Note that the points scored may depend on what is on route as well as on the estimated pleasure value of arriving somewhere. For example, arriving in Hamburg from Milan is worth much less than arriving there from Paris, because it is only the latter route that allows our student to see his friend again. But then, Milan as a city tempts him more than Paris, so the initial scores from Barcelona are in favour of Milan.

What to do overall? Tricky question. But remember the principle of dynamic optimization: *it does not matter how you got to Turku, you just think about the best way forward from there.* In the case of Turku (Fig. 5.2),

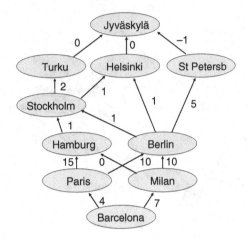

Fig. 5.2 A schematic presentation of the same routes as in Fig. 5.1, with the 'happiness points' added. Each number summarizes information on the pleasantness of the journey as well as the joy of visiting the city in which the arrow ends.

there is only one sensible option for the future action. This is a direct train to Jyväskylä, which is what we choose *assuming* we ended up in Turku at some stage in our lives. For future use, it is good to score what is the expected *future* happiness gain if one is in Turku. This is the value of the arrow of our chosen (only) option, i.e. from Turku to Jyväskylä, 0 points. So Turku scores the value 0 for the future happiness it offers. Helsinki likewise scores 0, and St Petersburg − 1, for the expected hassle of crossing the border. Note that this value does not yet include the happiness of arriving in St Petersburg, enabling the student to see all the sights. This will be dealt with below.

Now let's move backwards one step. So far it is not guaranteed that the final route takes our student through Stockholm, but assuming he might end up there, it is good to know what one should do from then on. There are now two options: the ferry to Turku (worth 2 points) or to Helsinki (1 point). He chooses Turku, but not strictly speaking because the ferry is more fun that way but because the *sum* of the current ferry trip and the future happiness provided by these two cities is greater for Turku (2 + 0) than for Helsinki (1 + 0). Stockholm scores 2 for the total of future happiness it provides, which is calculated assuming one behaves optimally from Stockholm onwards, that is, chooses the Turku rather than the Helsinki option. Indeed, we now know that going from Stockholm to Helsinki can never be part of an optimal strategy in this procedure, so we

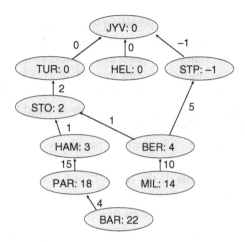

Fig. 5.3 The solution to the route choice problem. Arrows that are not part of optimal decisions have been removed. Each city has been assigned a value that tells the expected happiness points when following the optimal route from this city *onwards*. Note that this value refers to future gains and thus does not include the happiness already gained by visiting this city.

can remove the Stockholm–Helsinki arrow completely. Why? Because of the principle of dynamic optimization: no matter what route we might take to get to Stockholm, the future rewards gained by travelling to either Turku or Helsinki are fixed: they are 2 and 1, respectively. This is also the reason why one calculates *backwards*, from destinations near the end towards the beginning. Once the future route from one city (= state of being) onwards is fixed, we can assign a single expected value to the future rewards associated with being in this state. If one tried the same thing *forwards*, one would not be able to decide the best choice until all possible future route combinations have been tried. While this could still be doable with a small number of routes (Fig. 5.1), increasing the options would quickly make it impossible.

And so on: we move backwards in the graph, each time asking ourselves, *if* I ended up in a city X, what would I do from then on, given that I have already calculated the optimal routes from the next cities onwards? Scoring cities according to the future happiness points that moving on from a city provides, and removing arrows that won't be used, one finally arrives at Fig. 5.3.

The optimal route can now be read from the graph: from Barcelona to Paris, then to Hamburg with the friend, then to Stockholm, take the ferry to Turku, and then the train to Jyväskylä; total score 22 happiness points.

Note that the 22 points is the expectation one gains when adhering to the
solutions that maximize the long-term benefits, in this case, the duration
of the whole journey. Immediate rewards are not maximized (e.g. one has
to skip the idea of seeing Milan) and the arrival at latitude 60° ends up
being more boring (Turku) than the preferred alternative (St Petersburg)
because some other parts of the journey had more weight. Also, note that
the procedure has also automatically produced a 'plan B' for whatever
deviation occurs. In Fig. 5.3, there is no arrow that leads to Milan
because arriving there is not part of the optimal route. But there is an
arrow from Milan onwards. If the student for some strange reason ends
up taking the wrong train in Barcelona and finds himself in Milan, this
arrow tells the best option from then onwards: given this happened, it is
now best to adjust and travel the whole route in a completely different
way, going east via St Petersburg.

5.1 How to survive a winter day

In the example of travel through Europe, points were scored everywhere
along the way. In a biological setting, the analogy is, of course, fitness
that accumulates throughout the lifetime of an individual. For example, a
long-lived bird such as an albatross might breed every year and gain
lifetime reproductive success every time it produces a fledgling, but for an
albatross breeding is a huge effort. Trying to reproduce every year can
prove very costly in terms of survival. This is why reserving resources for
the future, and skipping some years between breeding attempts, may
prove better in the long run (Houston and McNamara 1999, p. 247).

In biological settings, it is almost always possible that the animal dies
before it reaches the defined 'destination' (the end of the time period
considered by the model). Sometimes, fitness is only accumulated if one
survives the whole length of the time period in question. This is not a
problem for dynamic optimization: points are then only scored if the final
destination is reached alive. Because individuals can rarely predict exactly
what will happen to them, many biological problems also require that
there are several possible final destinations, and that transitions between
states (the equivalent of cities in Fig. 5.2) can be probabilistic. This is not
a problem either, as we shall see.

Imagine that it is a winter day in the taiga forest. It is cold and quiet,
and after only a few hours of sunlight, dusk settles again. But nevertheless, a flock of small birds – tits (or in North America, chickadees),

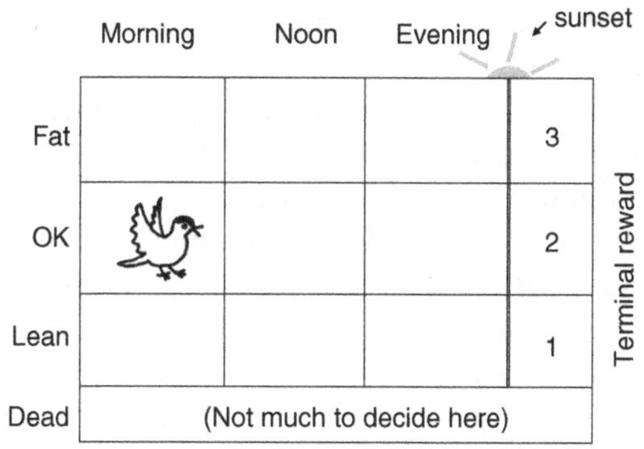

Fig. 5.4 The different states that a bird can find itself in: body condition can vary from 'dead' to 'fat', and time of day varies too.

perhaps treecreepers and some goldcrests – forages eagerly among the spruce trees. How do they manage? The sad truth is that some of them don't: winter mortality takes its toll. Life for a small bird in the winter is not easy, and each evening it is important to have enough energy reserves to survive a long, cold night. Add to this the problem that foraging itself is risky, and we have the basic outline of a problem. A bird can find itself in different states, determined by the time of the day (say, morning, midday and evening), and the body condition of the bird (fat, OK, lean, or dead; Fig. 5.4.)

In this case, there are no rewards to be gained except if the bird survives the whole day. The *terminal reward* at the end of the day (i.e. how many fitness points one scores when ending the day in one of the possible conditions; Fig. 5.4.) depends on how good the energy reserves are at that point. After all, we don't quite know how cold the night will be, and it is better to be prepared for the worst. So, while the terminal reward is positive for all birds that are alive, entering the night in a fat state is, say, three times better than in a lean state. Why three times? Strictly speaking, the rewards should correspond to the *reproductive value* of a bird at a specific point in time (McNamara and Houston 1986). This is a term of life history theory, summing up the expected genetic contribution of our individual to distant future generations (Houston and McNamara 1999). In our case, we don't, of course, want to tackle every possible life history question posed by small passerines. But no reproduction can occur in the future if the bird does not survive the winter, and we also know that fat stores are very essential for surviving a winter night. Hence we can assume that reproductive values in this context increase strongly with the fat stores achieved at the end of the day. (For guidance towards a better solution, see Section 5.2.)

Our task is now similar to the problem of the European traveller: for each state that the bird can find itself in, the best actions should be determined from now onwards. The states are tabulated in Fig. 5.4; the depicted bird finds itself in the state 'it is morning and my body condition is fairly OK'. So, what are the relevant actions? We assume that every time step (morning, noon, evening) the bird has two behavioural options: it can either forage or rest.

Foraging can be dangerous, as it involves easily visible physical activity, and active birds can fall prey to overwintering predators (Roth *et al.* 2006). To make matters worse, there is evidence that fat birds are less agile and cannot escape predators as well as lean birds (e.g. Witter *et al.* 1994; Metcalfe and Ure 1995; Kullberg *et al.* 1996; Lind *et al.* 1999; Krams 2002). Therefore, choosing to forage leads to the state 'death' with some probability that depends on the state of fat reserves of a bird. Let us denote the probability of death by d_i, where i is the bird's current condition (1 = lean, 2 = OK, 3 = fat). We can vary each value of d_i between 0 and 1, to investigate how predation risk is expected to influence foraging strategies. We can also investigate what happens if predation risk is made independent of the bird's body condition ($d_1 = d_2 = d_3$).

If the bird survives the foraging bout (which happens with probability $(1 - d_i)$), things look quite good: body condition improves. However, the speed of improvement could vary, for two distinct reasons. We could imagine that foraging in the winter is not the easiest task, and sometimes the bird fails to accumulate enough food for body condition to improve. Alternatively, we may judge that it is a too high rate to allow body condition to shoot up by one whole step (e.g. from 'lean' to 'OK') as a result of one successful foraging bout. Perhaps it is more realistic that an improvement of one condition step requires, on average, three foraging bouts. These considerations can be incorporated by introducing another parameter, foraging efficiency f, which is interpreted as the probability that a foraging bird (that is not caught by a predator) improves its condition by one step. For example, if we settle on an average of three foraging bouts for each condition step, we have $f = 1/3$.

A foraging bird, therefore, moves up the condition ladder with probability $(1 - d_i)f$, stays at the same condition with probability $(1 - d_i)(1 - f)$ and dies with probability d_i (Fig. 5.5a). It is worthwhile to perform a quick check to see that these probabilities sum up to 1: one of these mutually exclusive events must happen to the bird. Also, time always moves one

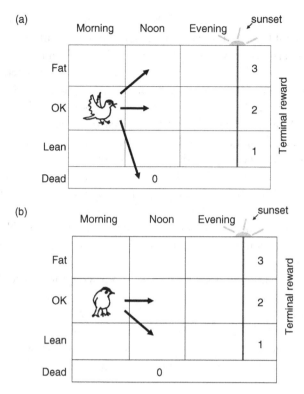

Fig. 5.5 The bird cannot decide precisely which arrow to follow; for example, predation events that lead to death are unpredictable. However, it can choose between scenarios (a) and (b). Thereafter, the probabilities of following each arrow are, from the topmost to the lowest arrow, for the foraging bird (a) $(1 - d_2)f$, $(1 - d_2)(1 - f)$ and d_2; for the resting bird (b), the probabilities are $(1 - c)$ and c.

step forward (Fig. 5.5a). And, of course, we will have to remember to deal with boundaries (e.g. a bird in top condition can no longer move up).

If the bird decides to rest, it follows a very different route. A resting bird will lose condition, and following similar arguments as above, we can define a probabilistic rate of condition loss c. A resting bird will move down the condition ladder with probability c, and retain its condition with probability $1 - c$ (Fig. 5.5b). Once again, the probabilities of the mutually exclusive alternatives sum up to 1.

Note that a bird can die for two different reasons before it reaches the terminal reward. It can be caught by a predator while foraging (probability d_i) or it can be lean and resting and lose too much condition and enter the death zone that way (probability c). Birds can, of course, die during the night too, but this is implicitly taken into account through giving them better terminal rewards if they are fat in the evening. One can, of course, complicate the model by assuming that predators sometimes find resting birds too. Many other complications are possible from food hoarding (Brodin 2000) to behaviour in dominance-structured winter flocks (Clark and Ekman 1995). Let's keep our first ever dynamic program as simple as we can, but a certain kind of change is possible without any extra modelling effort: we can make the number of time steps larger than three, and we can also include many more condition steps than just three. The discretization of the day into three parts (morning, daytime and evening), or that of bird states (lean, OK or fat), was done for annoying technical reasons: we have to be able to describe transitions between possible discrete states. The annoyance will be milder if there are very many states, which bears more resemblance to a continuous flow of time or a continuous distribution of different possibilities for body condition. The MATLAB program (Box 5.1) is built to deal with an arbitrary number of time and condition steps – we can choose to divide the day in to so many time units that they reflect foraging on a minute-by-minute basis, if we so wish. The program produces a table of optimal decisions in each possible state that the bird might find itself in, and a graphical representation of the strategy.

Box 5.1

Here we develop a function to calculate the optimal rules of foraging for a small bird on a winter day. Writing a program like this is an exercise in attention to technical detail. If condition 0 refers to the first row of a matrix, then condition 1 must be found at row 2, and a top-condition bird has condition maxc but its behaviour is not found looking at the maxcth row, but at the row maxc + 1. To constrain this kind of headache to the time of writing and checking the program, rather than repeating them every time the program is run, the program generates a helpful plot of the matrix in a way that makes it clear which actual values of condition or time a particular point in the matrix refers to. The aim is that blue and white squares mark foraging and resting, respectively. MATLAB provides a function pcolor that does something similar to what one wants, but it has a nasty habit of filling in averaged colours when two adjacent points have different values. So we have resorted to building our own custom-made graph. This program uses the MATLAB function fill to draw rectangles filled with a precisely defined colour at certain coordinates. The x and y coordinates of each rectangle are computed in such a way that, for example, the first row, second column reflects condition 0 along the y axis: the rectangle extends vertically from -0.5 to $+0.5$. Time step 2 (which is what the second column means) is indicated in that the rectangle extends horizontally from 1.5 to 2.5.

Consequently, some lines may look cryptic at first sight. The rules of responding to current condition are summarized in a matrix called ForagingRule. The first $+1$ when looking up values in the matrix ForagingRule(condition$+1$,t)$+1$, for example, takes care of referring to the correct row when thinking about condition, while the second $+1$ sorts out a small technical detail about choosing the correct colour. Colours are indicated as rows of numbers giving the intensity of red, green and blue. The resting decision is marked as 0 in the matrix, but to look up the relevant colour we cannot refer to the 0th row of a matrix that gives the colour. We have to tell MATLAB that if ForagingRule equals 0, the colour to be looked up is found in row 1 of colour, and if ForagingRule equals 1, the colour to be looked up is found in row 2. In general, we, therefore, have to add $+1$ to the value found in the matrix ForagingRule.

Tedious, yes (see Lesson 3 in the Appendix for general ideas about indexing), but once it is done, it will never have to be touched again. Often one finds that a particular part of a program could be useful more generally, and it may make sense to separate often-used procedures to become functions in their own right. This option will be used in Box 7.1 where a similar graphing procedure has become an independent function that creates 'pairwise invasibility plots' from any kind of fitness data it is given – even thoughtfully including an option for colour-blind people.

The function additionally makes use of truth values to decide if a particular condition is fulfilled (see Lesson 4 in the Appendix for details).

```
function ForageRule=forage(dmin,dmax,c,f,maxt,maxc)
% Forage=forage(dmin,dmax,c,f,maxt,maxc)
% dmax=probability of death per time unit if you're
% very heavy
% dmin=probability of death per time unit if you're
% very lean
% c=rate of consuming resources
% f=feeding efficiency
```

Box 5.1 cont.

```
% maxt = maximum time (= number of time units the day is
% divided into)
% maxc = maximum condition (= number of different
% condition units)
% The output is the ForageRule matrix, with 1 denoting
% foraging, and 0 denoting resting.

% Reminder: rows indicate condition, columns indicate
% time. Rows are chosen like this:
% dead = row 1, condition 1 = row 2, condition 2 = row 3,
% etc
% This means that best condition is maxc but this is
% found at row maxc + 1
% Terminal reward increases with condition
% so we already know the values for the last
% (i.e. maxt + 1st) row
Reward(:,maxt + 1)=(0:maxc)';

% then, probability of death increases linearly with
% body weight
d=[0 linspace(dmin,dmax,maxc)];

% anyone who is alive can either improve or then not ...
P_eat_up=(1-d)*f;
P_eat_same=(1-d)*(1-f);
P_eat_dead=d;
P_rest_same=1-c;
P_rest_down=c;

% ... except those who already are in top condition
% cannot improve so they get different values here
Ptop_eat_same=1-d(end);
Ptop_eat_dead=d(end);

% we start from the end of the day and continue
% backwards
for t=maxt:-1:1
  % individuals who are dead have index 1
  % individuals who are in top condition have
  % index maxc + 1
  % Rules for updating fitness values
  % first everyone except those who already are dead,
  % or in top condition
```

```
% We wish to compare two benefits: the expected reward
% from now onwards if one forages, and if one rests
for i=2:maxc
  RewardIfForage(i,t)=...
    P_eat_same(i)*Reward(i,t+1)+...
    P_eat_up(i)*Reward(i+1,t+1)+...
    P_eat_dead(i)*0;
  RewardIfRest(i,t)=...
    P_rest_same*Reward(i,t+1)+...
    P_rest_down*Reward(i-1,t+1);
end;

% Now the special cases
% dead ones don't get any rewards at all
RewardIfForage(1,t)=0;
RewardIfRest(1,t)=0;

% top ones can't improve their condition
RewardIfForage(maxc+1,t)=Ptop_eat_same*Reward
  (maxc+1,t+1)+Ptop_eat_dead*0;
RewardIfRest(maxc+1,t)=P_rest_same*Reward
  (maxc+1,t+1)+P_rest_down*Reward(maxc,t+1);

% Calculate the best foraging rule. This makes
% clever use of matrix notation as well as of truth
% values: if the statement is true, the value becomes 1,
% and zero otherwise. See Appendix, lessons 3 and 4.
ForageRule(:,t)=...
  RewardIfForage(:,t)>RewardIfRest(:,t);

% Update Reward by assuming individuals use the
% better of the two behavioural options in each case.
% The ~ means 'not', see lesson 4 in the Appendix
Reward(:,t)=ForageRule(:,t).*RewardIfForage
  (:,t)+~ForageRule(:,t).*RewardIfRest(:,t);
end;
% Now some graphical procedures. Each state is
% represented as a rectangle that will be coloured blue
% or white depending on whether one forages or not.
% Colours are defined as the intensity of red, green,
% blue
% The first row of 'colour' is pure white
% (1 unit red   +   1 unit green   +   1 unit blue)
% while the second colour is pure blue
% (zero units red or green   +   1 unit blue)
```

} one line

} one line

} one line

Box 5.1 cont.

```
colour=[1 1 1; 0 0 1];
% squares will need to extend to + -0.5 of their
% centrepoint, so the whole figure will extend to these
% dimensions
figure(1); clf;
axis([0.5 maxt+0.5 -0.5 maxc+0.5]);
hold on;

% This plots coloured squares in the correct position
% on a graph.
for t=1:maxt,
  for condition=0:maxc,
  % define the outline of the square in question
  xcoord=[t-0.5 t-0.5 t+0.5 t+0.5];
  ycoord=[condition-0.5 condition+0.5...
        condition+0.5 condition-0.5];
  fill(xcoord,ycoord,colour(
    ForageRule(condition+1,t)+1,:));
end;
end;
xlabel('Time');
ylabel('Condition');
hold off;
```

Solving for optimal behaviour follows exactly the same logic as the example of travel through Europe: start near the end and move backwards. This time, however, a bird cannot make a firm decision of the 'go to Turku' type (in this case, for example, 'improve condition'). It can only decide between the probabilistic outcomes that result from foraging or resting. Consider the example of Fig. 5.4 in which there are three categories for both time and body condition, and the terminal reward is assumed to increase linearly with condition. The first calculation of our backwards procedure concerns a bird in top condition in the evening: the top row, 'evening' column in Fig. 5.4. There is still some time left before the sun sets, so we need to know what the bird should do with it. If it forages, it can die (probability d_3, zero terminal reward), or it can stay at the same condition (probability $1 - d_3$, terminal reward 3). Remember, this bird cannot improve its condition: top-condition birds are a special case since there is no fatter state available. So, the expected future fitness for a bird foraging in top condition at the end of the day is $d_3 \cdot 0 + (1 - d_3) \cdot 3 = 3 - 3d_3$. What about resting? The bird can stay in the same condition with probability $1 - c$, reward 3, or drop one step, probability c, reward 2. Expected future fitness if resting equals $(1 - c) \cdot 3 + c \cdot 2 = 3 - c$. Now the crucial question. Which option yields a larger fitness expectation, foraging or resting? Resting gives a better expected reward if $3 - c > 3 - 3d_3$, which is equivalent to $c < 3d_3$. At small enough rates of condition deterioration, it is better to rest, the threshold being larger if foraging in top condition is particularly risky (i.e. the value of d_3 is large). This makes very good sense.

In practice, we will run the program with particular values for each parameter. The condition $c < 3d_3$ for resting will cease to apply for birds earlier in the day, because the expected fitness values associated with entering a particular state will be different: they depend on what is awaiting the birds in further future steps. The earlier back in time we move, the more complicated such calculations will be. But if we resort to single values of parameters at a time, we can note down the fitness associated with each state as a single number, making our task much easier. We might focus on the case of $c = d_3 = 0.1$, for example, and for the top-condition bird we will then find out that resting brings about fitness 2.9, and foraging 2.7. It is better to rest than forage, so we ignore the suboptimal 2.7, scribble 2.9 in the top right corner of the table, and also remember to note down that the optimal choice in this case is to rest.

This calculation is performed likewise for every other cell in the rightmost column. Then it is repeated for the preceding column, then the one before that, always moving backwards towards the morning. The result is a table with instructions on what to do in each possible state. It is good practice to do this once using nothing more than a pen and paper, but quite soon we will find it easier to turn to a computer – particularly as dynamic programming suffers from being a *numerical* procedure: to find out how a particular parameter, say c, influences solutions, we have to compute solutions for many different values of c.

The program (Box 5.1) makes the assumption that we wish to consider values of d_i that increase linearly between the lowest, denoted in the MATLAB program by dmin (probability that the leanest individual dies through predation), and highest value, dmax (death probability of the fattest). A choice of dmax > dmin will then reflect our assumption that being fat is bad news when encountering a predator. But we can also try out alternative assumptions, for example what happens if there is no predation at all: dmin = dmax = 0. As an initial choice, let us consider $c = 0.4, f = 0.8$. The day could be divided into, say, five time units, and the individuals are given six possible condition values in addition to being dead. Running the program (Box 5.1) then gives the following answer:

```
» forage(0,0,.4,.8,5,6)
ans =
```

0	0	0	0	0
1	1	1	1	1
1	1	1	1	1
1	1	1	1	1
1	1	1	1	1
1	1	1	1	1
1	1	1	1	1

OK ... now what does this mean? In the world of computing, 1 means 'true' and 0 means 'false', and the way we built our program, 'true' denotes 'yes, forage now'. Also, the indexing system of Box 5.1 works in such a way that the ith condition is given on row number $i + 1$; consequently, top condition is given on the lowest row. This is perhaps a little counter-intuitive, but indexing the rows the other way round would have been quite tedious too. To make sure that we will be able to interpret things intuitively the right way round, the program of Box 5.1 gives the result in a nice graphical form too, where top condition is plotted high up (Fig. 5.6a). The answer is clear: dead individuals (row 1) don't do anything, but everyone else forages (Fig. 5.6a). If there are no predators around, there is no harm in foraging as much as one can, so birds should always do it. Nice and intuitive.

Let us consider the effects of predation now. It probably makes sense to keep the predation risk fairly low, because it occurs every time step during which a bird forages. Values that are too high applied at every single time step will leave the bird with unrealistically small chances of surviving throughout a whole winter. So, let's increase dmax, but to no more than 0.01. Now the assumption is that leanest birds can always escape the sparrowhawk, and the risk increases linearly up to 0.01 for the fattest ones. The program now gives the output:

```
» forage(0,0.01,.4,.8,5,6)
ans =
```

0	0	0	0	0
1	1	1	1	1
1	1	1	1	1
1	1	1	1	1
1	1	1	1	1
0	1	1	1	1
0	0	0	1	1

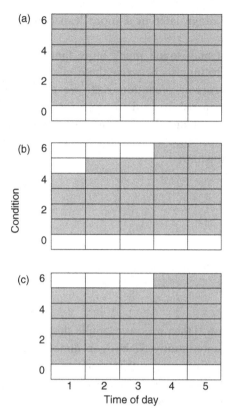

Fig. 5.6 Graphical representations of the solutions to the foraging problem, now with five time units in a day and six different values of body condition for live birds (additionally, a bird can also be dead). (a) There is no predation ($d_i = 0$ for all i), $c = 0.4$, $f = 0.8$. All birds that are alive forage continuously. (b) As (a), but predation risk increases linearly from $d_1 = 0$ (leanest birds) to $d_6 = 0.01$. Fat birds do not forage in the morning. (c) As (a) and (b), but the predation risk is now independent of body condition ($d_i = 0.01$ for all i). Fat birds still avoid dangerous foraging activities in the morning, but overall there is now less resting than in (b) – which is intriguing because foraging in (c) is overall more dangerous than in (b). See text for further explanation.

Now, the solution is quite intriguing (Fig. 5.6b). In the evening, everyone forages: it is good to approach the night in a fat condition. In the morning, however, fat individuals are quite happy to lose some weight. There may be two reasons. One, foraging is particularly risky for these individuals. Two, it may actually be beneficial to lose some weight: the evening is still far away, and it may be better for one's health to first lose weight and then forage later in a more lean condition, which makes foraging less risky.

The latter effect may sound much more fanciful compared with the former. Can we distinguish between these effects? If the matter is all about managing *current* predation risk, we should never expect an increase in predation risk to lead to more foraging. But this is exactly what happens if we change the previous parameter values such that *all* individuals now experience a predation risk of 0.01, regardless of their weight. Let's try this out:

```
» forage(0.01,0.01,.4,.8,5,6)
ans =
0    0    0    0    0
1    1    1    1    1
1    1    1    1    1
1    1    1    1    1
1    1    1    1    1
1    1    1    1    1
0    0    0    1    1
```

The slight difference to the previous example shows the power of dynamic programming. Counterintuitively, birds in almost top condition (row 6, column 1) switched from resting (Fig. 5.6b) to foraging (Fig. 5.6c), even though the change of parameter values means that foraging became riskier for them. The reason must therefore be that resting birds in Fig. 5.6b were doing it, at least partly, to lose weight adaptively and for a short time period only; the end reward is still higher for fatter birds in the evening.

What happens if predation risk is higher still? What happens if the rates c or f change? Can predation ever become so dangerous that birds stop foraging completely? Try first to generate predictions based on verbal reasoning, and then see if the program output matches your intuition.

5.2 Got interested?

For a great textbook on dynamic optimization, with plenty of examples, see Clark and Mangel (2000). A shorter introduction can be found in Ch. 5 of Mangel (2006). Several chapters (e.g. Chs. 6 and 7) in Houston and McNamara (1999) also describe problems that can be tackled using dynamic optimization, including ones where pay-offs depend on the actions of others, such that the dynamic (time-dependent) nature of the problem combines with game-theoretic considerations.

In our small-bird example, finding out the optimal strategy is only one answer that a dynamic program can give us. The obvious next question is how many individuals will actually be observed foraging or resting? The optimal rule might dictate that some birds do not forage at certain times, but observing this requires that birds actually enter the state that dictates this particular state-dependent behaviour. How many birds will be in top condition at time step 2, for example? If you pause to think for a while, you will realize that the answer depends on the number of birds that started the day in condition 1, 2, 3 and so on. It is a good exercise to write the program that calculates the number of birds that end up in states 0 (death), 1, 2, ... and top condition if the initial number of birds in each morning state is known. In this case, of course, we follow the probabilities *forwards*. Bookkeeping is much easier forwards than backwards, so if you have followed the backwards calculation, doing the same forwards is less daunting than it first seems. If using MATLAB, the first step is to form a vector of the number of birds in states 0, 1, 2, ... top condition and give it a name. The vector only gives the starting conditions, so it can contain any situation you wish to examine: for example [2 100 0 0 0 0]' indicates that there are two dead birds in the morning, 100 birds whose condition is low, and that there are four better options for condition but no bird actually exists in these states in the morning. (The ' at the end of the vector makes it vertical, so that rows correspond to condition like in the forage program). Then, for each state, check ForageRule to see what each of these birds do. And then? We already have the rules that specify what happens to birds when they forage or rest. Let's say that the value we have chosen as d_1 is 0.01, f is 0.5 and ForageRule tells us that birds in low condition forage rather than rest. The interpretation is that 1% of the 100 birds will die, and of the remaining ones, 50% remain in the same condition, and 50% improve. The column of expected bird numbers in the next time step must therefore become [3 49.5 49.5 0 0 0]': the birds who were already dead stay dead, while the 100 previously alive birds get redistributed among the states. The programmer's task is to write down the rules for this calculation.

And what about the whole winter? For this, we need some extra assumptions that we did not make explicit so far: how likely it is that a bird in, say, top condition survives the night, and what is its body condition the next morning? Considering this 'closes the loop' between the end of one day and the beginning of the next. We then no longer have to make hand-waving arguments about the importance of fat to decide what the terminal rewards should be. Instead, we now have precise equations that link earlier states (evening on day i) to future ones (morning on day $i+1$), and rewards in the evening must be based on what is expected to happen in the future. Since days follow each other, this becomes an iterative problem where reproductive values in one time step are calculated based on ones in the future. The necessary procedure is described in much greater detail in Ch. 9 of Houston and McNamara (1999; see also Mangel *et al.* 1994; McNamara *et al.* 1998). The cycle length, of course, does not have to be a day. For example, moulting feathers is costly, but it has to be done at some time during the year; a bird's reproductive value over time can then be modelled based on its current condition and whether it is breeding, moulting or migrating (Houston and McNamara 1999).

Models of overwintering birds have been used to illustrate the concept of reproductive values (McNamara and Houston 1986). Newer models have included a variety of factors that we did not consider (for a review see Brodin (2007)), including variable predation risk (Pravosudov and Lucas 2001), digestive constraints (Bednekoff and Houston 1994), the social dynamics of a flock (which necessitates a game-theoretic treatment (Clark and Ekman, 1995)), and the effects of hoarding (Brodin 2000; Pravosudov and Lucas 2001). Hoarding, obviously, is an alternative to body fat as a way to store energy reserves. A particularly interesting paper (Brodin 2000) discussed a dynamic model that at first produced exactly the wrong kind of predictions. One would think that hoarding species should gain fat later in the day than non-hoarding species, because they have a more predictable food supply and can rely on this stored food instead of carrying it dangerously 'around the waist' during the day. Simply dynamic programming procedures confirm this prediction, yet data on mass gain in wintering birds shows the opposite pattern: hoarders gain relatively more fat reserves in the morning than nonhoarders.

However, Brodin also found that weakening the relationship between fat stores and predation risk could switch the model to produce predictions that matched with field data. The lesson: do not feel too smug about your first model ('if the model doesn't fit the world, change the world … ') but be prepared to accept that some assumptions do not capture reality as well as they should. In the process, both the modeller and the empiricist will learn something: if assumption x led to a wrong prediction but y works much better in producing the observed pattern z, it is important to go out and see if conditions x or y prevail where z is found. This is how science progresses – and it is fun!

6

Game theory

where we get caught in a traffic jam,
and end up wondering where all those trees came from.

I live inside an area that is sometimes, scornfully, called 'traffic jam Finland' by Finns who live in less crammed regions of the country. Nevertheless, the situation is internationally speaking not particularly bad, mainly because of Helsinki's efficient public transport system. This does not (yet?) combine with road tolls, but these are certainly being discussed. A leading figure of the local green party, Osmo Soininvaara, recently[1] made the remark that driving from Puotila – an eastern district of the city – to the city centre could take as long as 45 minutes during the morning rush hour in the late 1960s. The city has grown significantly since then, but nowadays Puotila dwellers can expect to reach the centre in about 20 minutes, whether driving or taking the metro. Jams became a memory of the past when bus lanes were introduced, followed by a metro line in 1982.

Interestingly, drivers argued as vehemently against the bus lanes as they argue against road tolls now. Soininvaara's point is to remind us that in the long run it may pay to accept a personal cost to achieve a common good: if efficient public transport leads to less congested roads, this is obviously a benefit too for those who prefer to drive their cars. Therefore, Soininvaara suggested that the cost of driving should be shifted from taxing car ownership to charging more for the use of cars; everyone would eventually benefit from such an action. Why Soininvaara faces a tough challenge in convincing the dwellers of Helsinki, though, is a problem that is deeply rooted in *game theory*.

---

[1] Helsingin Uutiset, 22 September 2004.

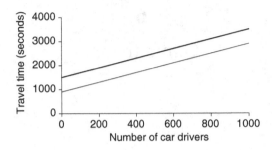

Fig. 6.1 Travel time when taking the bus (thick solid line), or when driving a car (thin solid line), and there are no bus lanes. In MATLAB, this figure can be created very quickly: x=0:1000; plot(x, 25*60+2*x, x, 15*60+2*x).

Game theory is designed to handle questions where the pay-offs (expressed as, for example, the ease with which one commutes to work) of an action depend not only on a player's own choices but also of those of others. Let us stop at a roadside to see how this works.

Imagine a situation where 1000 people have to use the same road each morning to get to work. They can each choose to use their own cars or a bus. Because buses stop for other passengers, and because one has to walk to the bus stop first and wait until the bus arrives, the travel time is always longer by bus. For example, we might note that if everyone travels by bus, the travel time is on average 25 minutes. Somebody then tries using his own car and notes that he can make the same journey in 15 minutes.

The more cars there are on roads, however, the slower the flow of traffic. Let's assume that every passenger who switches from using the bus to using his or her own car slows down other traffic (both cars and buses) by 2 seconds. What will happen?

Figure 6.1 shows the outcome. When there are no car drivers, it is very tempting to switch from using the bus to using one's own car. After all, it reduces the travel time from 1500 seconds (25 minutes) to 900 seconds (15 minutes). As more people switch, the travel time increases for both buses and cars. But no matter how many drivers there are, the temptation to switch remains the same: cars are always 10 minutes faster than buses. This makes the number of car drivers increase until all 1000 people use cars (right end of the figure). The situation is bizarre: now it takes everyone 2900 seconds (i.e. close to 50 minutes) to get to work – far longer than it originally took by bus!

The outcome is quite typical of game theory. Individuals cannot directly change the behaviour of others, but they can change the behaviour of

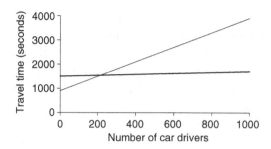

Fig. 6.2 Travel time as in Fig. 6.1, but with bus lanes. In Matlab:
x=0:1000; plot(x, 25*60+0.2*x, x, 15*60+3*x).

themselves. If they respond optimally to the current situation, the eventual outcome may well become suboptimal for the population as a whole. In the social sciences and in economics, the behaviour of our commuters is, oddly enough, described as *rational*. Rationality in this context means nothing more fancy than that every individual chooses the best action, given the current situation (e.g. one with 200 drivers and 800 bus passengers) and the currency in which benefits are measured (here, in terms of minimizing travel time). Considerations of the type 'what would happen if everyone behaved like me' are not part of this definition of rationality, and, as Fig. 6.1 shows, rational decision making does not always lead to particularly intelligent outcomes. If you think this is an exaggerated example, consider that the speed of traffic in central London in the year 2002, just prior to the introduction of the fees ('congestion charges') for drivers to enter central London, had fallen to a level equal to the days of horse carriages.

How to make the grumpy people caught in the traffic jam of Fig. 6.1 happy again? It is hard to change the fact that people will keep choosing whatever suits them best: no public campaign marvelling at the efficiency of public transport will work if reality shows that it is slow and inconvenient – regardless of whose fault it might be. So, let's try something different, and improve the efficiency of buses. The idea is simple: bus lanes.

In Fig. 6.2, the road has been changed such that buses have their own lanes for almost the whole of the journey. As a consequence, a passenger who switches from using the bus to driving now only slows down buses by 0.2 seconds, but slows down other cars by 3 seconds. Note that drivers disturb each other more than in the previous example, because cars now have only one lane available to them. The slopes of travelling time are now drastically different for buses and cars. What does a 'rational' (i.e. selfish) commuter do?

If there are no car drivers, the car lane is pretty empty, and cars still offer the same time advantage as before: 15 versus 25 minutes. Thus, driving becomes more popular. However, when one third of commuters have switched to driving, the car lanes are already congested enough that taking the bus is equally quick. If there are still more drivers, driving is so slow that taking the bus is actually quicker. As a result, we expect a flow back to using buses, and there is, biologically speaking, 'frequency-dependent selection' that favours bus use if there are many drivers, and car use if there are many bus users. The outcome is that approximately one fifth of commuters will use cars, the rest will use buses, and the travel time stabilizes at roughly 26 minutes – which is only minutely worse than the original travel time by bus, if people all had altruistically rejected car use. This is a *Nash equilibrium* of game theory: it describes a behaviour that, once adopted, is stable in the sense that it does not pay for any participant of the game to deviate from his or her current behaviour. In other words, since a single commuter now cannot get to work in anything less than 26 minutes, no matter what her choice of vehicle, we do not expect the system to deviate from this equilibrium.

The moral of the story is that human beings are quite good at behaving opportunistically, maximizing their own benefits – and this can often cause problems. But by changing the pay-offs of the game, one can (at least sometimes) guide the outcomes towards solutions that are better in the long run (Penn 2003). Much of policy making is exploiting the rationality of people: harder taxes on polluting industries, for example, creates economic incentives to switch to less polluting technology, which tends to work much better than relying on pure goodwill.

In applications of game theory in the social sciences, however, it is tricky to find out what people are actually optimizing. It is not always money. In our traffic jam example, we assumed it was travel time, but this can't be the full truth either. The attractiveness of public transport will also depend on its reliability, how safe and clean the journey is perceived to be, and the ticket system. With tickets, it is not just the price, but convenience too: there are cities where one is shouted at if one does not have exact change, and others where one swiftly waves an electronic pass...And, perhaps, some factors cannot be explained with any blend of selfish preferences: it is not unheard of that someone makes environmentally friendly decisions, at least when this does not incur too high personal costs.

The life of a modeller is not made easier by the fact that people often go for the most convenient option available, but sometimes they pay a lot of money to get a chance to exercise. In games directly related to bio-logical questions, things are simpler, at least in principle. When considering problems that involve reproductive success, there is a well-defined concept that natural selection is expected to increase. This is fitness, i.e. the expected genetic contribution of an individual to future generations. While there are some pitfalls with this concept too (see p. 89), it is far easier to identify the fitness consequences of, say, sex allocation rules in insects than to find out why exactly one works so hard to be able to afford a flashier car than one's neighbour (Frank 1999; Kahneman *et al.* 2006). In practice, game theory has been applied to evolutionary pro-blems both when the pay-offs reflect offspring production exactly and when they are proxies for fitness that are somewhat further removed from the ultimate measures of offspring production. This chapter presents an example that makes use of a proxy, while Ch. 7 develops an example where the link between behaviour and the contribution to future gen-erations is somewhat tighter.

6.1 A game theory model of tree growth

Let us leave the roadside and venture into a nearby forest. This is a much more calming environment. Listening to the hum of the wind in the tops of the tall trees sounds comforting, and there is little to suggest that the existence of trees may be based on stringent competition. But leaning against a tall trunk, we may start to ponder what the trunk is for. A lot of biomass just to push the leaves higher up. Why?

Thinking about it in terms of simple optimality does not lead us very far. To a first approximation, a plant is a machine that transforms sun-light and inorganic material into biomass. To achieve this in the best possible way, one might expect a flattish layer that captures as many sun's rays as possible. Leaves indeed are often very flat, but why grow so tall, just to push all of them a little higher up? To be closer to the sun? A quick calculation is enough to reject this idea: even the tallest trees reach only 100 m above ground (Koch *et al.* 2004), managing to reduce the distance between the earth and the sun (150 million km) by less than one billionth.

But growing taller than others can give a plant a distinct advantage. Plants shade each other, and anyone who has ever entered a tropical rainforest can tell just how dark it can be in the shade. As such, it is perhaps not important to be close to the sun – but it can be very important to be closer to the sun than one's neighbour. The situation resembles the joke in which two poor runners are chased by a tiger and one claims they will never be able to outrun the beast, to which the other one replies 'I don't have to outrun the tiger, I only have to outrun *you*'. Relative performance is what matters here. In other words: your pay-offs depend on what others do – which sounds very much like game theory.

Already in the early 1980s, when game theory had only recently been imported from economics and mathematics to biology, Thomas Givnish presented the following argument to explain the existence of tall plants (Givnish 1982). His argument was based on an analysis of leaf height of herbs, but we'll adopt a simplified approach here. Imagine a plant that experiences a trade-off between the amount of structural tissue (the stem) and leaf tissue. The plant's fitness is determined by two factors: firstly, how much leaf tissue there is in the plant (let us call this f), and, secondly, how much each leaf is able to photosynthesize (we will call this g). The amount of leaf tissue, f, can be expressed as the fraction of the plant's above-ground biomass that is not 'wasted' on nonphotosynthesizing tissue such as the stem. Leaf tissue is important because a plant's photosynthesis depends completely on this tissue, and everything else such as building good roots for survival, or forming seeds or fruit for reproduction, is ultimately limited by the amount of photosynthesis performed by the leaves. We are not interested in the details of plant reproduction here, but make the simple assumption that the more photosynthesis, the more there will eventually be seeds or pollen too. Therefore, for our argument here, we can consider the plant's fitness to be proportional to the product fg.

The next step is to define all possible growth strategies and look at their effects on the amount of leaf tissue f and the per-leaf photosynthesis rate g. A 'strategy' is simply 'what the plant does': one possible strategy is to allocate no resources to growing a stem, while another is to grow a long stem and leave nothing for leaf production. The latter may not sound like good advice for a plant, but it is important to include it so that all possibilities are covered. Next, we must think about the plant's strategy. This could be called h (for 'height'), with the value 1 denoting some absolute maximum ever achievable: for example, trees cannot grow much taller than 100 m because they ultimately depend on capillary action for getting water to the leaves, and gravity increasingly hinders this process at great vertical distances (Koch *et al.* 2004).

So, f and g should be functions of h. But what kind of functions? To begin with f, how should the amount of leaf tissue f depend on the plant's height? The taller the plant, the lower the proportion of 'useful' tissue, f.

Not that the stem is totally useless, of course – it may be absolutely required to raise the leaves out of the shadow. But be cautious here and do not count things twice; in our definitions the height benefit should appear in g, not f. Therefore, avoid the temptation to argue that f will ultimately be bigger if the plant has a high g because then it will produce more biomass and can grow larger in all dimensions ... Remember to advance one step at a time. Here f captures the costly aspect of plant growth only, hence f should always decrease with plant height h.

Do we have any biological clues as to what the shape of the function $f(h)$ should look like? Have a look at the trees around you. A photograph of a tall, old tree will not look the same as a younger conspecific, even if it is reproduced to appear the same size. The old tree has a much more solid and wide trunk. The laws of physics dictate that large things require more massive proportions to stay upright and mechanically stable than small things. An elephant has much more robust legs than an ant magnified to the same size. This phenomenon, called allometric scaling, already caught the attention of one of the world's first experimental scientists, Galileo Galilei. The laws of physics apply to plants equally well as to animals; thus, when a plant allocates more of its biomass to height (as opposed to growing wider), it must allocate a disproportionally large fraction of its biomass to structural tissue. This means that the proportion of leaf tissue, f, as a function of plant height h, reduces more steeply at high values of h. It makes sense, therefore, that $f(h)$ follows a shape such as that shown in Fig. 6.3a: the functions decrease faster close to $h = 1$. A good family of functions that accelerate their rate of decrease in this way is given[2] by $f(h) = 1 - h^{\alpha}$, with $\alpha > 1$. Here 'family' simply refers to the fact that there are several functions of this kind, one for each value of α.

Why this particular family of functions? It is a relatively easy one to deal with, but the results here will not qualitatively depend on the particular choice for the function. 'No qualitative difference' is a fancy way

---

[2] How did we choose this? How easily one finds convenient mathematical formulations for 'shapes' depends entirely on how much experience one has with functions, so I recommend playing with them a lot. One starting point is x^2, which many remember to be a nice concave curve: in MATLAB, try x=0:0.01:1; y=x.^2; plot(x,y). Another similar one is x^3; try plotting this and see how the shape changes. Now replace x with h, which we desire to have on the x axis, and multiply by -1 to get perhaps the simplest convex functions $-h^2$ and $-h^3$. These have nice shapes, but need to be shifted upwards so that $f(0) = 1$ and $f(1) = 0$: zero height means that there is no stem biomass, giving the possibility for 100% leaf tissue (hence the $f(0) = 1$), and maximum height ($h = 1$) should leave no leaves, hence $f(1) = 0$. Since $-0^2 = 0$ and $-1^2 = -1$, we notice that h^2 is only a constant short of the desired function. Adding the constant 1 does the trick, and for some generality we may try out different exponents and not just 2 and 3.

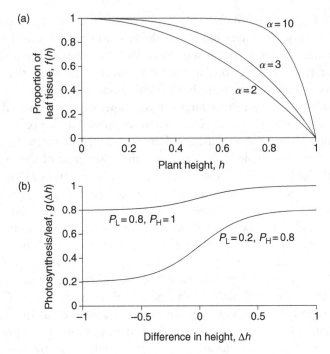

Fig. 6.3 Allometric scaling in plants. (a) The function $f(h) = 1 - h^{\alpha}$ relates plant height to the amount of leaf tissue, for three values of α. (b) The function $g(\Delta h)$ relates the difference in height between two plants to the photosynthesizing ability of the focal plant. The focal plant is the taller one if $\Delta h > 0$. The lower curve is derived with $P_L = 0.2$, $P_H = 0.8$ and the upper curve with $P_L = 0.8$, $P_H = 1$.

of saying that the exact numbers will vary but the main conclusions stay the same. We will show later why this is the case, so for the moment, do not worry. Figure 6.3a shows examples for different values of α. Inspecting this figure gives a biological interpretation for the parameter α: small values mean that growing taller has immediate large costs for leaf biomass, while large values allow the plant to grow quite tall quite easily before the full cost kicks in. For example, a plant species whose structural tissue is mechanically very stable will need less of it, and it will have a high value for α as a consequence. Consequently α could be called a kind of 'efficiency' or 'stability' of the structural tissue. Regardless of the value of α, however, there are no resources left for leaves to grow when h reaches the maximum value $h = 1$, which is exactly what we wanted.

It is perhaps good to stop for a moment and realize the limitations of the young versus old tree analogue. Young trees are good for showing that one can stay upright with a slim stem when one is small, but the function in Fig. 6.3a has strategies rather than age on its x (or rather, h) axis. This means that $f(h)$ is not meant to apply directly to young trees growing older and gaining height, instead it depicts the maximum proportion of leaf tissue for adult plants when they have already completed a growth strategy that gives a plant an eventual height h. The fixed biomass assumption also justifies our use of 'the total amount of leaf tissue' and 'the proportion of above-ground biomass that is leaf tissue' in an unusually exchangeable way: they equal each other if we denote the plant's total above-ground biomass as 1. (Question for the interested: in what way would the graph change if we applied the model in a different way, such that the total biomass was smaller for lower plant heights h? This interpretation would be appropriate for a plant that still grows.)

Next, we must decide the shape of g. This must be more complicated than f because photosynthesis depends on the amount of light falling on a leaf, which, in turn, depends on how high the leaf is situated relative to the leaves of other plants around. Other plants? Now we notice that we have not thought at all about how many there are, and how this might influence the situation. Indeed, in Ch. 7 we will spend much more effort thinking about the numbers of competitors. For the moment, however, let us follow the simplification that Givnish made too: the amount of light depends on the height difference between the focal plant, h_1, and its 'neighbour', h_2, limiting the number of players to two. The amount of sunlight for the focal plant can then be made to depend on the height difference between the two plants, $g(\Delta h) = g(h_1 - h_2)$. Here, the 'triangle' Δ is the capital Greek letter delta. It is a symbol often used to denote a difference, so we have introduced the notation $\Delta h = h_1 - h_2$ to mark the height difference between two plants. The notations $g(\Delta h)$ and $g(h_1 - h_2)$ are interchangeable.

What should $g(\Delta h)$ look like? We could start with a very strict view of the world of shading, where the taller plant gets all the sunlight, leaving none for the lower one. This would mean $g(\Delta h) = 0$ whenever $\Delta h < 0$, no matter how small the difference. However, I can think of at least four reasons why this is unrealistically drastic. Leaves are not solid bricks, instead they let some light pass through. There are also gaps between leaves, giving another route for sunlight to reach lower leaves. The plants do not grow exactly on the same spot but only next to each other, and finally, leaves usually do not form a perfect umbrella-shaped layer, but there may be vertical structure with some leaves higher than others. The last fact in particular means that h may be best seen as something that increases the *average* height of a leaf, and we can, therefore, expect a gradual improvement of $g(h_1 - h_2)$ as the focal plant grows taller (higher h_1), giving a higher proportion of leaves the advantage of direct sunlight.

As argued by Givnish, an S-shaped curve will probably work best for $g(\Delta h)$ (Fig. 6.3b). When two plants are equally tall, we have $\Delta h = 0$; therefore, $g(0)$ gives the amount of sunlight for one of two equal competitors. The curve should not increase without bounds when Δh becomes ever larger. When the difference in height is very large, growing taller no longer improves the amount of light falling onto the focal plant. Likewise, the curve should saturate to a low level for very low (and negative) values of Δh, where $h_1 < h_2$ and the focal plant is completely in the shade: it matters little if a dwarfed plant reaches a height of 5 or 10 cm next to a giant tree. A nice S-shaped curve is given by

$$g(\Delta h) = P_L + \frac{P_H - P_L}{1 + \exp(-5\Delta h)} \tag{6.1}$$

This may look baffling, but in fact we have encountered the beast before: it is once again the S-shaped logistic curve that resembles logistic growth of a population (p. 77). We already recycled it in Fig. 4.6 to depict a relationship between condition and survival. In our current formulation, we once again use parameters and constants to modify it to our current needs. The population dynamic version is $N(t) = K/[1 + (K - N_0)/N_0 \exp(-rt)]$ (see p. 77; or, for example, Case (2000)), and our Eq. (6.1) is exactly equivalent to this if we make the choices $r = 5$, $t = h$, $P_L = 0$, $P_H = K$ and $N_0 = K/2$. Our notation differs because we are not interested in population growth, but we nevertheless like the S-shape of the function: it captures the idea that relative height differences matter most when the two plants are roughly equal in height. Our '$\exp(-5\Delta h)$' ensures biologically sensible behaviour of the function when Δh is very small or very large. In the case of $\Delta h \ll 0$ (much below zero, i.e. the focal plant is suffering a lot from the presence of a taller neighbour), then $\exp(-5\Delta h)$ becomes a very large positive number, and $g(\Delta h)$ becomes

$$P_L + \frac{P_H - P_L}{1 + \text{something very big}}$$

which is the same as

$$P_L + \text{something very small}$$

In more precise mathematical language, $g(\Delta h)$ asymptotically approaches P_L as Δh tends to minus infinity. The word to look up in mathematics textbooks is a 'limit': no matter how small the focal plant, it will always be able to photosynthesize at least the amount P_L (and now it is clear why we chose P_L as a notation: 'photosynthesis at low light levels'). We can, of course, choose to set $P_L = 0$, if absolute, complete shading is possible.

Now if $\Delta h \gg 0$, then $\exp(-5\Delta h)$ is very close to 0 and $g(\Delta h)$ is now very close to $P_L + [(P_H - P_L)/(1 + 0)]$, which equals P_H. Therefore, the maximum amount of photosynthesis possible is P_H, and this is reached by being very, very much taller than the neighbour. Figure 6.3b shows two different choices for P_L and P_H. They could reflect different leaf structures: the less shade, the more light there is for everyone, and both P_L and P_H increase. However P_L probably increases more, as assumed in Fig. 6.3b. (Why?) In case you were wondering about the constant 5: without it the function would be S-shaped too, but not in a very pronounced way in the range of Δh values that we are interested in. We will return to the value of this constant in Section 6.2.

Now, the rules of the game have been set, and it is time to start striving towards a solution. Many introductory treatments of game theory present tables of discrete strategies as their starting point, with a list of two or more behavioural options: say, an animal could opt to 'fight' or 'not to fight' in a given situation. It is instructive to start with a glance at a small table here too, although we are immediately aware that this will not always give us the whole picture. There are, after all, more options for a plant than grow maximally tall or not develop a stem at all. (In reality, there are probably more options for an animal in a territorial contest too.) To include a little more variety, we will begin by giving each plant four options: no stem ($h = 0$), some investment in height ($h = 1/3$), a high investment in height ($h = 2/3$) or maximum height so that the plant consists of stem only ($h = 1$).

We are now ready to draw the table known as the pay-off matrix. Table 6.1 presents an example, calculated (Box 6.1) using particular choices $\alpha = 3$, $P_L = 0.25$ and $P_H = 1$, which means that a plant which is completely shaded by its neighbour receives 25% of the light that it would gain if the roles were swapped. Giving the two plants imaginative names such as plant A and plant B, each cell in the pay-off matrix (Table 6.1) requires two numbers: one for the fitness of A, the other one for B. For example, if plant A uses growth strategy $h = 1/3$ and plant B does not develop a stem at all ($h = 0$), the fitness for plant A equals $f(1/3)$ $g(1/3)$. The result is given as '0.848, 0.369'. In Table 6.1, plant A has fitness 0.848, while plant B has to do with 0.369 if it follows the strategy of not growing a stem at all.[3] Note that in this example plant B is the smaller one, so the difference between its height and that of its neighbour is negative where B's fitness is calculated (i.e. when B is the focal plant). The table also predicts – correctly – that if a plant has no leaves at all, its fitness is zero.

Table 6.1. *The pay-off matrix for two plants with four growth options[a]*

Height of plant A	Height of plant B			
	0	1/3	2/3	1
0	0.625, 0.625	0.369, 0.848	0.276, 0.686	0.255, 0
1/3	0.848, 0.369	0.602, 0.602	0.356, 0.620	0.266, 0
2/3	0.685, 0.276	0.620, 0.356	0.440, 0.440	0.260, 0
1	0, 0.255	0, 0.266	0, 0.260	0, 0

[a] Height growth options are $h = 0, 1/3, 2/3$ and 1; $\alpha = 3$, $P_L = 0.25$ and $P_H = 1$ (see text for details). In each cell, the first number gives the pay-off to plant A, and the second number the pay-off to plant B.

Box 6.1

A function used to calculate Fig. 6.4a (below), or to produce the table of payoffs. The program makes use of 'for' loops (see Lesson 5 in the Appendix), calculating fitness separately for each strategy pair that A and B might be using, among the options given in hvalues. This works fine for a few values, but gets very slow when the number of combinations to be checked grows. While 'for' loops appear more intuitive at first sight, Box 6.2 will give a more advanced version that runs much faster.

```
function plantheight(hvalues,alpha,PL,PH)
% function plantheight(hvalues,alpha,PL,PH)
% This computes a plot of best response curves for the
% two plants
% hvalues should give all height allocation options
% considered
% alpha is the exponent in f(h) =1-h^alpha
```

---

[3] Do not worry that these values are < 1: it does not mean that the plants will eventually go extinct. Our pay-offs are proxies for fitness, and we simply assume that all other things being equal, more photosynthesis as a whole is better for the plant. This allows us to ignore the step of translating these into expected reproductive success. See Ch. 7 for a model in which fitness is self-consistently derived.

```
% PL is the low limit of photosynthesis (poor light)
% PH is the high limit of photosynthesis (plenty of
% light)
% Examples: plantheight([0  1/3  2/3  1],2,.25,1)
% or plantheight(0:0.01:1,2,.25,1)
for i=1:length(hvalues)
  for j=1:length(hvalues)
    % Inside these loops, we investigate the option in
    % which plant A 'tries out' strategy number i, i.e.
    % hvalues(i) and plant B 'tries out' strategy number
    % j, i.e. hvalues(j)
    fA=1-hvalues(i).^alpha;
    fB=1-hvalues(j).^alpha;
    gA=PL+(PH-PL)./(1+exp(-5*(hvalues(i)-hvalues
      (j)))));
    gB=PL+(PH-PL)./(1+exp(-5*(hvalues(j)-hvalues
      (i)))));
    fitA(i,j)=fA*gA;
    fitB(i,j)=fB*gB;
  end
end
% if hvalues has very many values, we assume that one
% does not want to see the values rolling endlessly on
% the screen. Therefore we only ask to see them if hvalues
% is not a too long vector
if length(hvalues)<=15
  fitA
  fitB
end
% Now find the best solutions using 'max', and plot the
% result.
% Note the counterintuitive way to plot hvalues against
% A, and B against hvalues. This is because the
% definition of the 'x axis' differs in these plots
% between plant A and plant B, so for plant B the hvalues
% will go to what Matlab knows as the y axis
[tmp,bestindex]=max(fitA);
  bestA=hvalues(bestindex);
[tmp,bestindex]=max(fitB');
  bestB=hvalues(bestindex);
figure(1); plot(hvalues,bestA,'r.-',bestB,hvalues,
  'g.-'); grid
```

}one line

}one line

}one line

Nice! Next we will have to try to deduce the *evolutionarily stable strategy* (ESS). This is a central concept of evolutionary game theory, and it is very close to the Nash equilibrium. A 'strategy' is a rule that an individual obeys; in our case, for example, 'grow a stem of height 1/3' is a strategy. Strategies can be much more complicated, however, and the same strategy can produce different behavioural *tactics* in different scenarios: a strategy can, for example, dictate how tall to grow in good or poor lighting conditions, or how a barnacle responds to the scent of a predator (see Ch. 3). Two different competing strategies in such a case could differ in the threshold value that makes the barnacle make use one of the 'tactics' of straight or bent development. A strategy is *evolutionarily stable* if it cannot be challenged by any other strategy: that is, if using any other strategy leads to lower fitness.[4]

Despite the language of 'strategies', we do not of course require that our poor plants are capable of any rational or strategic thinking. Evolutionary strategies are typically expected to be genetically coded. In Table 6.1, there is a conflict between plants over reaching enough light, but the real battleground is over the gene frequencies that code for growing to a particular height. If a genotype makes the individual obey a strategy that constantly leads to low fitness in competition with others, we will not find such genotypes in future generations.

The usual jargon is to call the alternative challenges *mutants* that try to *invade* a population. While this may sound a bit like a plot of a very bad science fiction film, it simply refers to a scenario where a population is obeying a rule – say, individuals typically do not grow much of a stem – and we ask if an alternative gene will spread if it first arises in small frequencies (through mutation) and makes an individual behave differently, for example grow taller than the rule in the population (the '*mutant strategy*' differs from the '*resident strategy*'). If the mutated gene confers a fitness advantage, it will spread, and the original strategy will have turned out to be *invadable*. But if we have a strategy that cannot be invaded by any conceivable mutant strategy, it is evolutionarily stable: an ESS.

---

[4] The exact mathematical definition also allows for the possibility that there are alternative strategies that have equal, rather than lower, fitness compared with the one being challenged, but if individuals using an alternative strategy play against *each other*, they have lower fitness than when the individuals using the ESS play against the alternative. This prevents their spread in the population. If we denote by $w_{p,q}$ the fitness of p when playing against q, the ESS conditions are that either $w_{p,p} > w_{p,q}$ or, if $w_{p,p} = w_{p,q}$, $w_{p,q} > w_{q,q}$. These conditions guarantee that mutants q are selected against when rare; see, for example, Maynard Smith (1982), Bulmer (1994, p. 154) or Mesterton-Gibbons (2000, p. 75).

Back to our plants reaching for light: can we find an evolutionary stable equilibrium? Let us start with the top left corner '0.625, 0.625'. What if plant A deviated from its 'choice' of not investing in a stem and instead carried a mutant allele dictating it to allocate $1/3$ of its biomass to growing a stem? The plant will enjoy a fitness advantage: even though this plant will have less leaf tissue since $f(1/3)$ is smaller than $f(0)$, outcompeting plant B for light more then compensates for the loss. Photosynthesis at $g(1/3)$ is so much larger than at $g(0)$ that the fitness for A increases from 0.625 to 0.848 when A switches from $h=0$ to $h=1/3$, and B's strategy remains unchanged (1st column of Table 6.1). So, the strategy pair $\{0,0\}$ in which neither plant grows a stem cannot be evolutionarily stable, and we do not even need to check if plant B benefits similarly from altering its strategy. Which, by the way, it does: check the second numbers of each cell in the first row to verify this.

What about the strategy pair $\{1/3, 0\}$, in which plant A grows a little bit of a stem and plant B stays low (Table 6.1, second row, first column)? For plant A, this is indeed an excellent choice: for this individual, choosing another height would only give lower fitness values (check the first numbers of the first column), but now plant B begs to differ: if this plant carries a mutant allele that makes it adopt the strategy $1/3$ too, B's fitness will increase from 0.369 to 0.602. Hence, $\{1/3, 0\}$ cannot be evolutionarily stable either.

And so on, keep checking, always remembering that plant A is not assumed to be able to change the genes present in plant B. This means that selection can only act on A's strategy if *vertical* deviations in the table give a fitness benefit to this plant, and B can only be selected to change its strategy if it can be done using *horizontal* deviations. The first case, where vertical deviations by A would only deteriorate A's fitness and horizontal deviations by B would only deteriorate B's fitness, is the case where both use the height strategy $h = 2/3$. Here is our ESS.

But we should not stop yet. Sometimes *multiple* equilibria can be found that are all evolutionarily stable. In our traffic example, we only found a single equilibrium for the number of car and bus users in each case. However, the system we studied might have been more complicated: bus companies have a tendency to make their services very infrequent if there are few passengers, and to make buses run frequently if they are used a lot. In such a case, the convenience of using a bus can increase dramatically with the number of bus passengers as waiting times become shorter. The result can be described with two alternative equilibria: in one city there is little temptation to use cars because everybody uses buses and this makes maintaining good bus services easy. In another city, nobody uses buses and this gives no incentive for anyone to improve the services either. There may be nothing initially different in the economics of bus or car use, but either type of vehicle choice creates positive feedback towards heavier use of that particular system. Thus, frequency dependence of the pay-offs sometimes means that initial conditions (even totally random fluctuations in the number of passengers, or some other factor that initially has only small effects on passenger numbers) can completely determine which culture predominates later,[5] and each of them can be stable against 'mutant passengers' who try to benefit by behaving differently from the norm.

Because there could be multiple equilibria, we have to check all combinations of possible plant heights. It turns out that evolutionary stability is not found anywhere else in Table 6.1 than at {2/3, 2/3}, and – at least if plants are only given the choice among the four options in Table 6.1 – we predict that the evolutionary endpoint is 'grow a stem of 2/3 units tall'. The performance of each plant, in terms of total amount of photosynthesis achieved (our fitness proxy), is 0.440. Now, have a look at the possible fitness values and notice something strange. If both plants had 'agreed' not to waste resources on structural tissue, both would have been much more productive (fitness 0.625 for both) than at the ESS! There is conflict between the plants in the same way as there was conflict between passengers and car drivers in Fig. 6.1: the selfish act (use the car, or outgrow the other individual) can reduce the overall well being of all individuals in the group. Natural selection does not seem to equip plants with any better mechanisms to avoid pointless outcomes than those our commuters experience.

A handy graphical way to depict evolutionary stability is to ask what the *best response* to a certain strategy is. For example, in Table 6.1, if A uses $h = 0$, the best response of B is the one that gives it the highest fitness values among the second numbers in row: B should use $h = 1/3$. If A uses $h = 1/3$, the best response by B is again to outgrow A, now by choosing $h = 2/3$. If A uses $h = 2/3$, B's best response remains at $h = 2/3$, while if A uses $h = 1$, B's best response falls again to $h = 1/3$. The whole procedure can then be performed again, swapping the roles and looking for A's best response to B's actions. We get similar (symmetrical) best-response curves. Figure 6.4a shows a summary. The direction in which the graph is read is swapped between the two individuals: the horizontal axis serves as the 'x axis' when determining B's best response to how A is growing (solid lines, open dots),

---

[5] Support for this theory can be readily obtained by travelling in different countries.

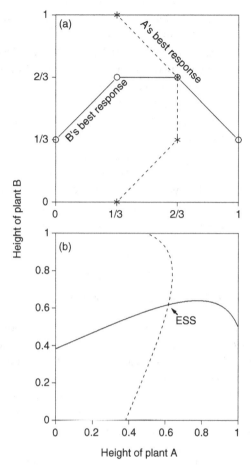

Fig. 6.4 The best responses of plants to each other's heights when $\alpha = 3$, $P_L = 0.25$, $P_H = 1$ and (a) plants have only four different height options, or (b) plants can choose any value between 0 and 1 (in practice, this is approximated with 501 different options between 0 and 1). Solid lines or open circles refer to B responding to A's height, dashed lines or stars show how A responds best to B's height, the latter being given on the y axis. In MATLAB, (a) is produced with the command `plantheight([0 1/3 2/3 1],3,.25,1)`, while (b) can be produced with either `plantheight (linspace(0,1,501),3,.25,1)` or with `plantheight_advanced ([501 501],[3 3],.25,1,0)`; the latter runs very much faster.

while the vertical axis should be used as the 'x axis' that gives B's behaviour to which A is responding (dashed lines, stars). The lines cross at a point where each individual's best response to the given situation is a behaviour that maintains the situation: this is where the ESS is found.

Now, remember that we knew already from the start that it is quite restrictive to assume that plants can only choose between four different height values. The jagged appearance of Fig. (6.4a) is, indeed, created by our initial refusal to give the plants more than four height options. With a computer at hand, the best responses can be quickly drawn for a much larger number of different growth possibilities, essentially forming a continuum between $h = 0$ and $h = 1$. The result is a much smoother picture of what is going on (Fig. 6.4b). Up to a point, it is best to grow taller if one's neighbour is tall, but trying to outcompete ridiculously tall neighbours (ones that can hardly grow any leaves) will not pay off. The best responses remain below 0.65, and now that a more precise determination of the ESS is possible, we find it at $h^* = 0.62$ (the * is often used to denote evolutionary stability). Our initial, coarse guess of $h^* = 2/3$ was not terribly far off.

6.2 An analytical solution?

This section is for those who like to wonder if it is possible to keep catching salmon (Ch. 2): can we reach an analytical solution here? The answer is yes, to some extent at least. Consider that mutants can always invade plant A's current strategy if there is a direction, either towards smaller or larger h, in which A's fitness increases (while B's strategy is kept constant). This increase can be quantified as $\partial(f(h_1)g(h_1-h_2))/\partial h_1$, which is notation that means 'take the partial derivative of $f(h_1)g(h_1-h_2)$ with respect to h_1': or, in plain English, $\partial(f(h_1)g(h_1-h_2))/\partial h_1$ is the slope of A's fitness against h_1. If this is not zero, h_1 cannot possibly be the best thing for A to do (at least if we are not dealing with a boundary case, see p. 73–74), and thus cannot form an ESS. Differentiating a beast such as $f(h_1)g(h_1-h_2)$ uses the product differentiation rule, and yields $f'(h_1)g(h_1-h_2) + f(h_1)g'(h_1-h_2)$. Here, the ' is yet another way[6] to denote differentiation with respect to h_1, handy when aiming at compact expressions. We now know that an ESS must satisfy $f'(h_1)g(h_1-h_2) + f(h_1)g'(h_1-h_2) = 0$. Rearranging, we get the condition

---

[6] Derivatives really seem to attract alternative notations and expressions. 'Differentiate' means the same as 'take the derivative', and possible notations include $\partial f/\partial h_1$, df/dh_1 (when there are no other variables that could confuse), $f'(h_1)$, and others too – derivatives with respect to time sometimes have a dot above the name of the function. The mess has been there ever since Isaac Newton and Gottfried Wilhelm Leibniz were busy discovering the rules of calculus, which led to a bitter dispute about priority. It comes as no surprise that both Newton and Leibniz invented and preferred to use their own notation.

$$-\frac{f(h_1)}{f'(h_1)} = \frac{g(h_1 - h_2)}{g'(h_1 - h_2)} \tag{6.2}$$

The left-hand side of this equation is positive, because it is doubly negative: firstly because of the minus sign and, secondly, because $f'(h_1)$ is negative, indicating a decreasing function $f(h_1)$. Equation (6.2) is an example of what is called an *implicit solution*: it states a condition that has to be fulfilled for something to be true, but it is not expressed in a way that would allow us to see how the evolutionarily stable value of h depends on the values for α or P_L or P_H in our model.

In fact, this abstractness is a strength too. Equation (6.2) does not depend at all on the particular choices we have made about the shapes $f(h)$ and $g(\Delta h)$. We can forget about the values of α or P_L or P_H and use this general form to deduce, for example, what would happen if we chose a completely different shape for $g(\Delta h)$. After all, our initial choices were quite restrictive: for example, the factor 5 in Eq. (6.1) is there only because it produces a beautiful, smooth but not too slow transition from low to high light levels. But there could be, for example, faster transitions, and it would be nice to know what happens then. Of course, one could make this factor another parameter of the model, and try out values of 6, 7, 8, perhaps 15 or 100 ... But with the general power of Eq. (6.2), we do not even have to constrain ourselves to functions that contain, say, exponential terms. Instead, we can now ask what happens to *any* potential solution if $g(\Delta h)$ becomes steeper. For example what if we have plants that have much broader leaves? Miraculously, Eq. (6.2) can be used to draw general conclusions without saying much more about the f and g than whether they increase or decrease with h.

Here's how. A steeper gradient in light levels according to relative height, or a steeper response of photosynthesis rate to current light levels, has the effect of making $g'(h_1 - h_2)$ larger in Eq. (6.2). Therefore, the right-hand side of Eq. (6.2) is now smaller than before and the equation as a whole is no longer satisfied; consequently a new value of h_1 must be found. To play its proper role in being part of an equilibrium, any candidate value for h_1 must produce a smaller left-hand side than before. Now since $f(h_1)$ is a decreasing function of h_1, the numerator $f(h_1)$ will decrease when h_1 is larger. A larger h_1 will also lead to a larger absolute value of the denominator $f'(h_1)$, as long as we assume that plants suffer progressively more in terms of leaf tissue when height keeps increasing. A smaller numerator and a larger absolute value of the denominator is exactly what we need to make the left-hand side smaller. Therefore, a higher value for h_1 will help us to satisfy Eq. (6.2) again, while a lower value for h_1 will never do (the arguments are similar). We're getting close to a conclusion that if there are two different kinds of condition leading to two different equilibria, the conditions in which shading has more severe consequences for photosynthesis (steeper g') will, all other things being equal, promote taller growth – *no matter what exact shape we chose for our functions*. All that is required from the functions is that $g(\Delta h)$ should be increasing, and $f(h)$ should not only decrease but accelerate its rate of doing so with increasing h. (The functions should be differentiable too.)

It still should be checked, of course, that an equilibrium value of h actually exists that satisfies the new requirements of Eq. (6.2), and that the effect of $g(h_1 - h_2)$ does not destroy our conclusions, since so far we checked the effects of a changed g' on f' and f, but not on g. Can a change in g destroy our argument? Not really, as long as we are interested in symmetrical pay-offs, which lead to symmetrical best-response curves. At any equilibrium in which $h_1 = h_2$, $g(\Delta h)$ should be unchanged in the end. Therefore, we can pat ourselves on the back: well done, we can reach some very general conclusions here without the need to worry that our particular choices for the functions led to highly unusual behaviour. Similar logic can be used to deduce other very general properties of the model.

6.3 Variations on the growth theme

Now that we have a model, it is nice to play with it. Box 6.2 gives an improved version of the program compared with that in Box 6.1. It runs faster thanks to a technical trick called vectorization, and it has two added biological features. Firstly, the different plants (A and B) could obey different trade-offs, expressed as plant-specific values of α, P_L and P_H. By far the easiest situation to imagine is that A and B belong to different species; obviously their structural and leaf tissues could be built differently, while they can still play games against each other. The program in Box 6.2 asks for two different values of α, but P_L and P_H are still assumed to be identical for both plants (if you are keen, feel free to relax this assumption too). Secondly, the two plants could be close relatives. This is obviously not an option if the plants belong to different species, but it can be an interesting alternative to investigate. Why? Simple: because it can matter. Since Hamilton's seminal contributions (Hamilton 1964), we have known that sometimes we must think about *inclusive* fitness rather than just the individual's own fitness.

Box 6.2

The program in Box 6.1 works, but the 'for' loops make it very slow when aiming at a very accurate solution. Here is a solution that takes advantage of *vectorizing*: calculating many values at once, instead of painstakingly trudging through each fitA(i,j). The result is much neater and faster (see Lesson 5 in the Appendix). This version assumes we will always investigate a full range of *h* values from 0 to 1; a parameter *n* controls how many different values are investigated for plant A, n(1), or plant B, n(2). This version of the program also has two additional features: it gives us the option of investigating two different values of α for the two different plants, as well as looking at the effects of relatedness *r*.

A technical detail: we use MATLAB's 'meshgrid' feature to create many values to be investigated at the same time. This necessitates forming matrices H1 and H2 instead of just a vector hvalues, as we did in Box 6.1. The two matrices together will form all the possible combinations that the two plants can find themselves in. To remain as consistent as possible with Box 6.1, we ask meshgrid to produce H2 first, then H1. This is because meshgrid by default produces as its first output a matrix in which values change in the columns, and the second output has the changes along the rows. In Box 6.1, we decided that rows denote what plant A is doing – hence the 'swapped' notation in the current program too. In cases of doubt, it is good practice trying out some solutions with unequal values for n(1) and n(2), which forces the two plants to have a different number of height options. In such a case, the program will halt if we mistakenly program some illegal combinations, such as trying to plot the best response of A against A's own strategy, rather than against B's strategy – only the latter is correct.

```
function plantheight_advanced(n,alpha,PL,PH,r)
% function plantheight_advanced(n,alpha,PL,PH,r)
% This computes a plot of best response curves for the
% extended plant game
% n should be a 1x2 vector describing how many
% candidate strategies are examined for plant
% A(n(1)) and plant B(n(2))
% alpha should be a 1x2 vector, describing its value
% for plant A (alpha(1)) and plant B (alpha(1))
% PL is the low limit of photosynthesis (poor light)
```

Box 6.2 cont.

```
% PH is the high limit of photosynthesis (plenty of
% light)
% r is relatedness between the plants

% create vectors that describe options for A and B
h1=linspace(0,1,n(1)); h2=linspace(0,1,n(2));

% now create matrices H1 and H2 that contain all
% combinations of values required
[H2,H1]=meshgrid(h2,h1);
deltaH=H1-H2;

% now that everything is in matrix form, all
% subsequent calculations can be performed at once
% (without 'for' loops)
fA=1-H1.^alpha(1);
fB=1-H2.^alpha(2);
gA=PL+(PH-PL)./(1+exp(-5*deltaH));
gB=PL+(PH-PL)./(1+exp(+5*deltaH));
% question: why did the '-' become a '+' here?
fitA=fA.*gA+r*fB.*gB;
fitB=fB.*gB+r*fA.*gA;

% now find the best solutions
[tmp,bestindex]=max(fitA); bestA=h1(bestindex);
[tmp,bestindex]=max(fitB'); bestB=h2(bestindex);
figure(1); plot(h2,bestA,'r.-',bestB,h1,'g.-');
grid
```

But before turning our attention to relatedness, let us first have a look at changing other values of the parameters. Figure 6.5a,b depicts cases in which the photosynthesis depends on relative height steeply (i.e. plants have a strong shadowing effect on others) or less steeply, using examples for $g(\Delta h)$ as given in Fig. 6.3b. Figure 6.5c additionally shows the solution for a case where height does not influence light levels at all $(P_L = P_H = 1)$. The conclusion? If plants shadow each other very badly (e.g. by having very broad leaves; Fig. 6.5a), they should grow higher in an attempt to outcompete each other. If the shadowing effect is not so bad (Fig. 6.5b), investment in stems should be smaller too. And if there is

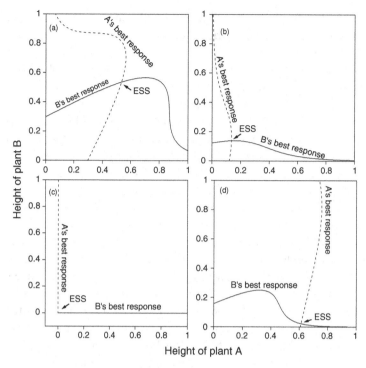

Fig. 6.5 (a–c) The best responses of plants to each other's heights when $\alpha = 2$, and (a) there is a strong shadowing effect, $P_L = 0.2$, $P_H = 0.8$; (b) the shadowing effect is weaker, $P_L = 0.8$, $P_H = 1$; or (c) the other plant's height does not influence the focal plant's photosynthesis at all, $P_L = P_H = 1$. In (d), the plants differ in their values of α, which is 10 for plant A but only 1.5 for plant B; the shadowing effect is moderate, $P_L = 0.5$, $P_H = 1$.

no shadowing effect – for example when the plants do not, in the end, grow close to each other, or if photosynthesis is limited by something else than light – the ESS is not to grow a stem at all (Fig. 6.5c). This means that the stem only represents an attempt to outgrow the other plant. It is curious that this attempt is both costly and unsuccessful. In each case (Fig. 6.5a–c), the plants are in the end equally tall, and because stems are costly to produce, the total amount of photosynthesis achieved will be less than it could be if neither plant invested in its stem. Quoting Falster and Westoby (2003), 'If ecosystems were truly cooperative organizations, vegetation would be a thin skin of green near the ground, without expenditure on stems detracting from the production. Not surprisingly, the most productive food systems on Earth (crops) comprise strategies with minimal investment in height.'

Cases with different trade-offs for the different species produce, perhaps unsurprisingly, asymmetric solutions in which one species grows much taller than the other. In Fig. 6.5d, it is far easier for plant A to grow tall ($\alpha = 10$) than for plant B ($\alpha = 1.5$). Accordingly, plant B does not try to outgrow plant A at the ESS but remains at a very low size. Plant B appears to reflect an alternative solution to the game of competing for light: stay close to the ground but develop lots of leaves to gather all of what little light is available there. Interestingly, several solutions in Fig. 6.5 show that plants are expected to 'give up' if their neighbours grow very tall: the best response to neighbours growing taller is initially to grow a taller stem too, but only up to a point; thereafter, the best way to deal with the competitive situation is to stop investing much in a stem, but invest in many photosynthesizing leaves instead.

How about the case with related plants? The key concept is inclusive fitness: an individual that follows Hamilton's rule '$br > c$' (Hamilton 1964) should do something that brings about a benefit b to a relative with relatedness coefficient r (e.g. 0.5 for a full sister) if the product br is larger than the cost to self, c. Hamilton's rule is often discussed in the context of altruism: in our case, a plant might not grow as tall as it otherwise would in order to avoid shading a relative. But how large are b and c? What is the baseline 'as it otherwise would', when in the absence of any competition the plant would not grow to any appreciable height at all (Fig. 6.5c)?

Difficult question – but there is a way out. Following Hamilton's rule maximizes inclusive fitness: this is the sum of all fitness consequences weighted by relatedness, self included. In the language of b, r and c: the inclusive fitness of doing 'something' equals $-c + br$, where the consequence for 'self' is $-c$ scaled by relatedness to self, which equals 1, and the consequence for the sister is b, scaled by relatedness r. If this sum is greater than the inclusive fitness obtained by doing something else (e.g. not being altruistic with no consequences to self, giving the sum 0), then one should do this something, whatever behavioural or life history strategy it was that we were thinking about.

In our case, the fitness of plant A using height strategy h_1 becomes $f(h_1) g(h_1 - h_2) + r f(h_2) g(h_2 - h_1)$, while that of plant B becomes $f(h_2) g(h_2 - h_1) + r f(h_1) g(h_1 - h_2)$. When $r = 0$, this reduces to fitness the way we were calculating it before. With $r > 0$, we get new solutions (Fig. 6.6) showing that plants, indeed, should compete less intensely for light if they grow with relatives. When the two plants are clones ($r = 1$), they will do best by abandoning the competitive effort of stem growth altogether (Fig. 6.6c).

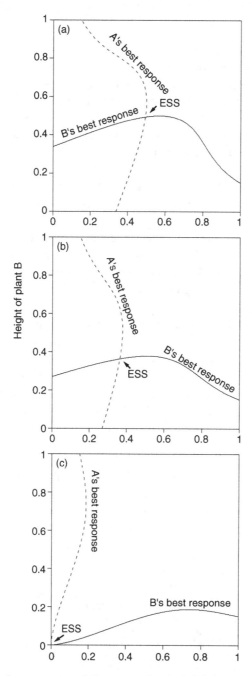

Fig. 6.6 The best responses of plants to each other's heights can also depend on relatedness, and high relatedness reduces stem growth at the evolutionary stable strategy(ESS) point. Parameters used are $\alpha = 3$, $P_L = 0.5$, $P_H = 1$, and (a) $r = 0$, (b) $r = 0.5$, (c) $r = 1$ (clonal plants). For example, (b) is created using `plantheight_advanced([501 501],[3 3],.5,1,0.5)`.

Do plants really grow altruistically[7] lower when surrounded by relatives? The idea of adaptive responses to avoid competing with kin is not new as such – examples range from dispersal to sex ratio biases – yet I am not aware of any direct tests of the plant growth hypothesis presented here (although shape-related fitness has been quantified in plants growing in groups, e.g. Aspi *et al.* (2003)). This shows how rewarding modelling can be: it really helps to form new ideas.

Some care is needed when using inclusive fitness arguments. Strictly speaking, they are approximations of what kin selection will favour (as is Hamilton's rule '$br > c$' itself), making use of some qualifying assumptions. For example, the rule relies on extrapolating the effects of small changes in an individual's behaviour, which is fine as long as selection is weak (Hamilton 1964). This can become a problem, however, if the rule is applied for discrete strategies with drastic fitness consequences, such as whether or not to kill a sibling.[8] More importantly still, one should not fall into the trap of double accounting. There are many ways to mess up fitness calculations. For example, if an individual does something to improve her sister's offspring production from three offspring to four, it is incorrect to give the sister an improved fitness score, add this weighted by 0.5 to the focal individual's score and still include *all* the four offspring weighted by 0.5 as part of the focal individual's fitness. Some care is required here, and the simplest way is to always count:

$$\text{inclusive fitness} = \text{own fitness} + r \cdot \text{others' fitness}$$

and then compare different inclusive fitness values for all the options of the individual's behaviour. This means that the differences that the focal individual's behaviour makes to her sister's fitness are all measured against the same baseline fitness. Not helping, for instance, still means that the

---

[7] This is, of course, only altruism in the sense that it gives an immediate advantage to another individual, not in the sense that it would do ultimate harm to one's own genetic contribution to future generations. If lower growth had the latter effect, it would not escape the grinding action of natural selection. This highlights the fact that evolutionary explanations of altruism necessarily work by 'explaining it away': what looked like true selflessness turns out to be altruistic only in the short term.

[8] The rule may also apply if selection is strong, but this is not guaranteed. A good example is siblicide: if we take a dichotomous view and consider alleles for killing and not killing the sibling, fitness consequences can obviously be large. Bulmer (1994, pp. 193–194) discussed a case where Hamilton's rule can be applied with ease if the older sib can kill the younger one, but equations become much more complicated if both can try to kill each other. Other good examples of the failure of Hamilton's rule in discrete games played between relatives can be found in Wenseleers (2006). For how to proceed in complicated settings that may include repeated interactions between kin individuals, see Taylor and Frank (1996), Day and Taylor (1997), and Taylor *et al.* (2007).

three offspring exist and should be included (weighted by 0.5); helping means that four offspring exist and should be included (weighted by 0.5); the difference that results from the behaviour is now correctly calculated.

If this appears too simple to worry about, consider that sometimes we need to think about overlapping generations where social animals can interact over prolonged periods of time, and each of them can vary their behaviour. Counting 'who gets what' then really requires care (Creel 1990a,b; Koenig and Walters 1999; Wolf and Wade 2001). For complicated scenarios, it is advisable to get familiar with the so-called direct fitness method, which offers notation that keeps one's head clear. The focus on 'direct fitness' means that the model keeps track of fitness at the individual level, but when deriving evolutionary predictions the method acknowledges that the fitness of one individual can covary with the fitness of others: Taylor (1996); Taylor and Frank (1996); Ch. 4 in Frank (1998); Ch. 7 in Rousset (2004); Lehmann and Keller (2006); and Taylor *et al.* (2007). The method also extends to causes of covariation other than kinship (e.g. membership in a social group (Gardner and West 2004)).

6.4 Got interested?

The model by Givnish, which is presented here in a simplified form, is by no means the last word on plant height games. There are probably many ways in which you can find it unsatisfactory. For example, how can the shade only depend on the plant's height difference, but not on how many leaves the taller plant has? Or, the two plants are simply assumed to exist at a fixed size rather than having to grow to their eventual size. One would, therefore, like to consider time-dependent dynamics as in Ch. 5, but the method presented there now appears insufficient because fitness in Ch. 5 did not depend on what others do. Can timing aspects be combined with the game-theoretic nature of competitive growth? Certainly it is possible. A recent review (Falster and Westoby 2003) discussed no fewer than 14 plant height games, and seven of them consider time strategies such as age and size at maturity. There are details to be sorted out, but the general message is quite robust: if competition with neighbours did not play a role, the most efficient way to organize plant material would be a thin flat layer of biomass. It is indeed an intriguing idea that all the world's magnificent forest ecosystems could be a result of an ultimately counterproductive and wasteful game.

Evolutionary game theory is, of course, not limited to the plant world. It is very widely applied in the study of animal behaviour. A concise summary of the state of the art is found in Nowak and Sigmund (2004), and an edited volume by Dugatkin and Reeve (1998) explores diverse examples ranging from the organization of social groups to games of sex ratios, habitat choice and foraging. No wonder; it is indeed a very common finding that fitness depends on others' actions, and a whole lot of mathematical theory is built to examine the consequences. Camerer (2003) provided a different focus, discussing game theory in the light of empirical studies of human behaviour. A useful starting point for further learning of the mathematical details is the book by Mesterton-Gibbons (2000); for a more concise summary see Ch. 8 in Bulmer (1994), Ch. 9 in Rice (2004) or Ch. 2 in McElreath and Boyd (2007). Houston and McNamara's book (1999) on models of adaptive behaviour contains excellent chapters on evolutionary game theory too, also extending the dynamic approach of Ch. 5 to game-theoretic considerations. The seminal volume by Maynard Smith (1982) is still worth a read too. Finally, Ch. 3 in McElreath and Boyd (2007) gives a beautiful explanation of a puzzling aspect of Hamilton's rule: how can it apply when siblings are said to share half of their genes but the vast majority of genes are known to be identical in humans and in chimps?

There is some squabble about how likely it is that natural selection will, indeed, produce an ESS. A fair summary of this discussion is, in my opinion, a 'cautious yes' (e.g. Eshel 1982; Parker and Maynard Smith 1990; Roff 1994; Gomulkiewitz 1998; Eshel and Feldman 2001; Rice 2004). The important point to consider is that developing a model always gives the modeller full power to choose what constraints the organisms can or cannot overcome. Obviously, if in reality there is no genetic, physiological or cognitive machinery in place to produce a strategy that is included in a model, it cannot evolve either (Parker and Maynard Smith 1990). Other factors can intervene too, such as very small population size (Orzack and Hines 2005).

Therefore, like all phenotypic models, the ESS approach assumes that evolution has had enough 'material' available to 'invent' the machinery necessary to produce the strategies that lead to victory. 'Material' here refers to sufficiently many individuals, subject to suitable mutational input, over a sufficiently long time (Eshel and Feldman 2001; pp. 293–295 in Rice 2004). In this sense, phenotypic models (optimization and game theory) can be seen to be better suited to finding evolutionary outcomes that will be generated over very long time periods – the 'what can exist' question, with a typical answer being 'a forest in which the height depends on factors x, y and z'. Models that are more directly linked to the genetic level tend to take existing genetic constraints as given, without giving the model a chance to overcome them. They, therefore, produce more accurate outcomes over shorter timescales – the 'what will happen right now' question, with an answer such as 'under current conditions there is selection for greater height' (Eshel and Feldman 2001; Pigliucci 2006). The longer-term evolutionary predictions may be less clear (Kotiaho, 2007). It is well worth reading Pigliucci and Schlichting (1997) in this context, and a clear discussion applied to a special case is found in Wilkins and Haig (2003).

However, stating this as a general difference would be an over-simplification. It is perfectly possible to build game theory models with a very rigid and limited set of behavioural options, and it is possible to build genetic models in which alleles can alter the overall genetic archi-tecture, performing all sorts of macroevolutionary miracles. It all depends on how the modeller chooses the constraints, and it is clear that this should be done wisely. Mutations that equipped zebras with machine guns would certainly give them an advantage when dealing with lion attacks, but the real-life relevance of such a model would be limited. In the case of plant growth, *Arabidopsis* plants have been shown to react to local density, growing taller when needed to avoid shade, but this kind of phenotypic plasticity again carries measurable costs (Weinig *et al.* 2006) – which offers another layer of complexity for the argument that games do not optimize population performance. But could a plant also sense if it grows next to related neighbours, or do we expect it to respond to average relatedness values only? Much of the value of any model is determined by the plausibility of the assumed constraints.

It also follows that theoreticians should think about how 'final' the endpoints of evolution are when the world in reality never reaches an unalterable equilibrium. An interesting special issue of the *Journal of Mathematical Biology* in 1996 (pp. 483–688) was devoted to reconciling the view of evolution as a continually changing process with the idea of stable equilibria (see also Eshel and Feldman 2001). For a highly readable summary of the ideas, see Marrow *et al.* (1996). For a philosopher's view on the timescale issues, see Rosales (2005) and, for a much more critical view, Sarkar (2005).

In Ch. 7 we shall have a look at an additional layer of complexity. In the current chapter, the number of participants in the game was fixed: 1000 in the commuting game, and two in the plant height game. In Ch. 7, fitness will not only depend on the actions of competitors but also on their *numbers*.

7

Self-consistent games and evolutionary
invasion analysis

where winter is approaching once again,
and we wonder if the promise of the coming spring
should convince us to stay put.

What does the title of the current chapter mean? The idea of self-consistency is simple: the model should not contain contradictions of logic. This sounds fairly obvious, but consider the following pitfall. In a model of parental care, a male is given a specific pay-off (fitness gain) if he deserts his brood after reproduction has taken place and starts searching for other females. If this pay-off is greater than the fitness gain from caring for his current brood, the strategy of deserting will spread. This is the essence of a very influential game of sex roles and parenting decisions (Maynard Smith 1977), yet current analysis (Houston and McNamara 2005) reveals that it is not self-consistent. Maynard Smith made the assumption that pay-offs from deserting depend on how the male's own mate responds to the situation, but another kind of frequency dependence was not included: the pay-off stayed constant, regardless of how common the strategy of deserting males was in the population. Why is this dodgy? Because if there is an increasing number of males searching for a limited number of females, the competitive situation must change as a result, and the pay-off should respond accordingly. Also, the number of females available must change with their breeding strategies. It is inconsistent to assume that males can gain heaps of offspring by deserting, if all females are busy caring for offspring and none of them is available to be fertilized. The general point is that, often, fitness depends not only on what the others do but also on *how many* of these 'others' are present in the

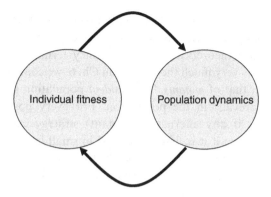

Fig. 7.1 The eco-evolutionary feedback loop.

population. In Ch. 6, this additional layer of complexity was irrelevant: we assumed that there are always 1000 commuters getting to work or that a tree either had a neighbour or alternatively (as a way to interpret Fig. 6.5c) it did not. But within a game, the number of neighbours was fixed. In the current chapter, we will consider cases where the total number of players – in other words, population dynamics – has an influence on our equations for fitness. This is very commonly the case: fitness is, after all, a measure of the demography of one's descendants, while population dynamics is a sum of demographic events. Hence we expect a close link between evolutionary and ecological outcomes (arrow from individuals to populations; Fig. 7.1). Conversely, the state of the population can have effects on what behaviours are selected for in individuals: whether a male cares for offspring or deserts the current brood should depend on the number of females that he is expected to encounter if he deserts. This, in turn, depends on the care and desertion decisions made by other females, while the level of competition that the focal male experiences will be influenced by how many other males have made the decision to desert and look for new females (Houston *et al.* 2005). The result: the back arrow in Fig. 7.1, which completes the eco-evolutionary feedback loop between evolutionary and population dynamics.

Models that make the link between game theory and population dynamics explicit make use of *evolutionary invasion analysis*. 'Adaptive dynamics' is another term that is often used to describe models that focus on the eco-evolutionary feedback.[1] Within this framework, abstract-sounding

---

[1] To be exact, adaptive dynamics is a subset of evolutionary invasion analysis where the temporal pattern of mutant introductions is explicitly modelled (see Ch. 12 in Otto and Day (2006)).

terms such as 'invasion fitness' or new abbreviations ('PIP' for pairwise invasibility plots) have been defined to allow researchers to communicate more easily about theoretical advances. Fancy terminology aside, the basic idea remains very much the same as in Ch. 6: we compare the fitness of *residents* with that of *mutants*. The *resident* population strategy is one that currently prevails in the population. For every possible resident strategy, we ask if any alternative (mutant) strategy can invade (i.e. increase its numbers) if it is first introduced in small quantities.

Typically, the average fitness of a resident equals one. This is because we focus on the ecological equilibrium: the population is regulated through density-dependence, which means that each individual, on average, replaces itself during its lifetime through reproduction. A mutant whose expected lifetime reproductive success (LRS) exceeds unity will then spread (not perhaps the first time the mutation arises but, given the advantage of this mutation, eventually it will). However, if this mutation spreads so that it becomes the new resident strategy, density-dependence kicks in again and, with the new strategy firmly in place, the population is again at ecological equilibrium such that, on average, LRS $= 1$. But now, a new and even better mutant could have LRS > 1, and it starts to spread and so on... Finally, we may be able to find a strategy that has LRS $= 1$ *and* cannot be invaded by any alternative, and so we have found evolutionary stability.

The only difference to the previous chapter is that this time we are required to spend a little additional effort first on calculating the *environment* that the resident strategy creates. The environment in this context includes any factors that change depending on what the individuals in the population are doing. In Ch. 6, the environment was simply the average height of a competitor that a tree experiences: this is the scenario that a mutant tree has to cope with. A spreading strategy would change the average height of competitors in a population. In this chapter, we consider that *numbers* of competitors can change too, and this should be included in the definition of the relevant environment.

7.1 Partial migration

Bird migration is one of the true wonders of the world: millions of individuals depart from their breeding sites every autumn, often travelling thousands of kilometres before reaching their wintering sites. The return journey, equally long, often ends precisely at the previous year's breeding site. As a result, birds can enjoy ecological niches that would otherwise be very hard to fill: insectivorous birds breeding in the tundra, for instance. Sometimes, however, birds appear to avoid migration in surprising environments. The wintering birds in Ch. 5 endure quite harsh conditions, and there are even insectivorous birds (e.g. goldcrests) that every year attempt survival in the taiga forest, looking for small overwintering insects amongst the spruce branches. Not every goldcrest stays put, however; in the northern part of their breeding range many migrate, which makes goldcrests one of the many *partial migrants* in the world.

Why do some birds leave and others stay? One hypothesis is that the nonmigrating birds take advantage of the *prior residence effect*: an individual who has arrived somewhere first seems to gain an ownership right that protects it quite strongly against takeover attempts. This effect is an empirical observation, widespread across taxa (Kokko *et al.* 2006b), examples ranging from sea urchins (Shulman 1990) and spiders (e.g. Moya-Laraño *et al.* 2002) to fish (e.g. Parmigiani *et al.* 1987) and mammals (e.g., Neumann 1999). Why exactly such respect for prior ownership should evolve is another good question (e.g. Kemp and Wiklund 2004), and one that deserves modelling effort of its own (Kokko *et al.* 2006b). But in the context of partial migration, let us take the effect as given and ask if it can explain partial migration.

If migration makes birds miss the opportunity to reserve the best breeding locations by overwintering, it might be best to stay put. But migrating could offer better survival prospects. Migrating through unfamiliar landscapes is, of course, not without its risks, but if migration did not offer a net survival benefit, it would be hard to explain why any bird would ever migrate. So, to understand partial migration, we will need to specify differing survival prospects, as well as the prospects of territory acquisition. But let us start by defining the strategy used in the population. It should have something to do with whether one migrates or not, so here it is: x, which we call a propensity to migrate. Each bird in the autumn departs from breeding sites in the autumn with probability x and tries to return the next spring. A fraction $1 - x$ of autumn birds, therefore, stay put. Allowing the birds to use any value of x (between 0 and 1, naturally) is a sensible choice in the same way as allowing the trees of Ch. 6 to use any height strategy between 0 and the maximum. We do not want to add unnecessary constraints, such as the requirement that a bird always does the same thing.

Modelling the difference in survival is not particularly hard. We can simply assume that such a difference exists by stating that a migrating bird survives the winter with probability s_m, whereas a nonmigrant[2] survives with probability s_n. But what about the prior residence effect, allowing better access to good habitats? This is much trickier, and much of the following pages are devoted to finding out what kinds of habitat the different individuals end up inhabiting. We can aim for as simple as possible a model by assuming a very strong prior residence effect: let's say that nonmigrant birds that have survived are allowed to reserve the very best breeding sites, and all migrants that return have to do with what is left. Still, what exactly is the quality of areas that migrants will return to, and hence their reproductive success?

What we want to find out is the reproductive success of migrants, which we could denote by R_m, and that of nonmigrants, denoted by R_n. But before even trying to find out what R_m and R_n are, let's consider what we will do with these values once we have them. A bird that migrates can expect to gain (on average) R_m offspring every year, and the number of years it can breed is .... Well, how many is this, if annual survival is s_m? If this sounds similar to the problem on p.79, you're entirely right: we need once again exactly the same infinite sum of probabilities of surviving one, two, three, ... breeding attempts. The answer is the same: the expected number of breeding events is $s_m/(1 - s_m)$, and hence the expected LRS of a bird that always migrates is $R_m s_m/(1 - s_m)$. The expected LRS of a bird that never migrates is calculated similarly, and equals $R_n s_n/(1 - s_n)$.

But, unfortunately, this time we cannot stop so easily. What about a bird that uses, say, $x = 0.5$ and tosses a coin each autumn to decide if it should migrate or not? Is its fitness equal to the mean of migrant and nonmigrant fitness? This is not obvious: after all, the bird should survive through a random sequence of migration and nonmigration years, and it should not be taken for granted that it all averages out in the simplest of ways (in fact, it does not). Therefore, rather than writing down possibly incorrect expressions for LRS, it is easier to use an alternative approach and to think of 1 year in a bird's life at a time. In a population where a

---

[2] Note that we shall avoid the use of 'stay resident' in this chapter to mean the nonmigrating strategy, because of the unfortunate and unavoidable collision with the term 'resident strategy'. The resident strategy describes anything that a population is assumed to be currently doing when a mutant is about to invade, and, therefore, the 'resident strategy' may mean, for example, that 90% of all individuals migrate! There is little one can do about this pitfall: game theoreticians did not develop their theory with partial migration in mind, nor did ornithologists check whether there might be a confusion arising with theoreticians when they were developing their terminology. In Ch. 8 we will encounter entomologists, whose 'migration' term is very different from what ornithologists mean by 'migration' – so theoreticians are not the only ones to blame. Once again, this is a lesson in how essential it is to be alert when listening to the jargon of various fields of science.

fraction x of birds migrates, one individual alive in the autumn, prior to autumn migration, will contribute

$$xs_m(R_m + 1) + (1 - x)s_n(R_n + 1) \tag{7.1}$$

individuals to the next autumn's population. A mutant performs better than the resident if its probability to migrate, which can be denoted by x', leads to a better growth rate compared with the growth of the resident population that still uses x. The growth rate of the mutant's strategy is called 'invasion fitness'. (Once again: do not get confused by the fact that x and x' had a totally different meaning in Ch. 2, and in Section 6.2 the $'$ denoted a derivative. We have once again encountered the versatility of mathematical notation – a major strength and also a major headache.)

One could think that our task is completed; all we need to do now is to consider R_m, s_m, R_n and s_n as parameters that get different values for different species or scenarios, and see what values of x are favoured... but no, this chapter is not that short!

Why not? Remember, we swept for a moment the problem under the carpet that we do not yet know R_m and R_n. It is fine to consider s_m and s_n parameters, which means that we will simply pick suitable values to investigate them, but R_m and R_n are different: they must depend on x, because the qualities of vacant territories cannot stay constant if the behaviour of the population, described by x, changes. For example, there will be more overwintering birds if x is low, and this influences the average breeding success of both nonmigrants and migrants, because of competition for breeding habitat. This also means that the reproductive success of migrants and nonmigrants can depend on the fraction x. That is, there may be frequency-dependent selection that can, perhaps, help us to understand how two different strategies can persist in the same population.

So, let's start thinking about how exactly R_m and R_n depend on x. This is something we cannot know, unless we specify what is available in the area as a whole: we need to know the quality distribution of breeding habitat. This, of course, could obey any kind of distribution. For example, it could be that every site is of equal quality but the number of sites is limited (e.g. a certain number of nest cavities or boxes in a homogeneous forest). Or, there could be two clearly different types of habitat: for example, oystercatcher breeding success on salt marshes depends heavily on whether there is direct access to mudflats or not (Ens *et al.* 1995). Or, site quality could be normally distributed. There are so many possibilities that it is hardly possible to consider all of them, even

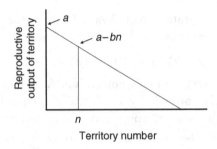

Fig. 7.2 The relationship between territory number (ranked according to quality) and reproductive output.

though sometimes mathematical arguments can be generalized for all types of distribution (for an example in a territorial context see Kokko and Sutherland (1998)).

Here, let us try to keep things simple. We will need a limited number of territories, and they should also differ in quality. The simplest choice is that reproductive success declines linearly from the best to the worst territories. Therefore, when territories are ranked according to quality, the nth territory yields expected reproductive success $a - bn$, where a and b are parameters (Fig. 7.2). Large values of a shift the line upwards and increase the reproductive success on all territories, as well as the number of territories that give positive reproductive success. Large values of b make the decline in quality with rank steeper and hence reduce average reproductive success in the habitat, as well as the number of territories available. To allow for a large population to exist, b should consequently be small.

Our assumptions mean that if a bird arrives when the n best territories are already taken, the bird has to do with a territory of quality $a - b(n + 1)$. Still, we don't know exactly what is the expected quality of breeding habitat of nonmigrant or migrant birds is, unless we know the total size of the population too. Why? Because all birds can find high-quality sites if there are very few birds in the population; while at larger population sizes, some will have to accept poorer sites, and the average breeding success will decline. At the population level, this is highly interesting as it creates population regulation (Rodenhouse *et al.* 1997; Gill *et al.* 2001). But it is also important for individual fitness, because the expected rewards of migrating versus overwintering will depend on the difference in breeding success that follows.

So, to find out how well migrants are doing, we need to know how many nonmigrant birds and how many returning migrants compete for territories in the spring. These numbers depend not only on survival and the fraction of individuals that use either way to survive the winter (i.e., the 'resident strategy' x), but also on how well individuals were, on average, reproducing in the previous breeding season – but because of site-dependent success this, in turn, depends on how many there were... Help! Too many things to consider! Panic! Impossible! Well, not quite. All we need to do is to

- calm down
- assume a particular resident strategy, which, remember, we express as the fraction x of birds that depart in the autumn
- ask what kind of equilibrium numbers we then predict for migrant and nonmigrant birds at a particular time of the year (e.g. springtime)
- then write down the fitness consequences of using a different strategy (i.e. a mutant value x').

Now, if a fraction x of birds departs in the autumn, what is the number of birds arriving in the spring? This is the competitive environment created by the resident strategy. Well, we don't know this yet (because we didn't know the total autumn population size), so let us denote the number of arriving spring migrants by n_m. This is not a parameter but a value that will be calculated. The number of nonmigrant birds that have survived their overwintering attempt is denoted by n_n, so the total spring population size equals $n_m + n_n$.

Since we assumed that nonmigrants get to choose territories first, and we also assume that they can accurately detect territory quality, they will obviously pick the n_n best sites. What is the average breeding success of a nonmigrant who has survived? Since we assumed a linear decline in success with the number of territories, it must be the average of a and $a - bn_n$. The per capita reproductive success of survived nonmigrants is, therefore (see Fig. 7.3)

$$R_n = 0.5\,a + 0.5\,(a - bn_n) = a - bn_n/2 \qquad (7.2a)$$

Then, migrants get their territories, occupying those territories that are left over. What is the quality of leftover territories? Have a look at Fig. 7.4. There we see that the leftover places must offer reproductive success that varies between $a - bn_n$ and $a - b(n_n + n_m)$. The average reproductive success of a survived migrant is, therefore, the mean of these two boundaries, which is

$$R_m = a - bn_n - bn_m/2. \qquad (7.2b)$$

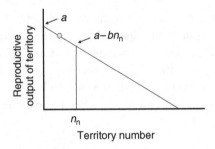

Fig. 7.3 The average reproductive success achieved by nonmigrants is the mean of a and $a - bn_n$, when there are n_n nonmigrants that have survived the winter.

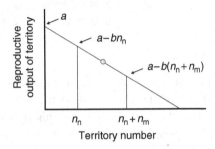

Fig. 7.4 The average reproductive success achieved by migrants is the mean of $a - bn_n$ and $a - b(n_n + n_m)$ when there are n_n nonmigrants and n_m migrants that have survived the winter.

There are two caveats to think about. Firstly, to be really exact, we should follow our definition of the relationship between territory number and quality, which assigns the first nonmigrant bird in the population a reproductive success of $a - b$, rather than a. Likewise, the first migrant should have reproductive success $a - b(n_n + 1)$ rather than $a - bn_n$. This influences the averages R_n and R_m a little bit, but the differences are vanishingly small if populations occupy large areas with many territories. The second, bigger, problem is that we have assumed that there are never more individuals than there are territories. If there are, some cannot reproduce at all, and this will have to be taken into account in the calculations of reproductive success. Such a scenario can arise if territory quality is high, leading to a 'surplus' of individuals being produced.

Because taking into account the nonbreeding possibility complicates the equations somewhat, this adjustment is left to the keen reader to make. As such, it is not difficult: think about all the possible cases and calculate mean reproduction taking into account that some individuals cannot breed at all. In the examples that follow, we have chosen to be somewhat more relaxed about this issue, and instead we only check if the problem arises with each set of parameter value. The problem does not appear if the equilibrium size of the total population, $n_n + n_m$, is still within the region where the latest-arriving migrants can still reproduce: that is, $a - b(n_n + n_m)$ is positive. If it is not, the program prints out a warning that the current parameter values deserve the extra work described above.

Let's move on now: we are still not finished with determining what sort of competitive environment the resident strategy x will create. How much will the population grow from one spring till the next? Let's first consider what happens during the summer. In other words, as a starting point let us pretend that we know the spring population sizes in a particular year t. The total number of individuals breeding in that year is $n_n(t) + n_m(t)$, which means that the autumn population size in the same year must equal $n_n(t) + n_m(t) + n_n(t)R_n(t) + n_m(t)R_m(t)$. The first two terms refer to the parent birds that are still present, and the last two terms to their offspring. The bracketed index t is needed to remind us that reproductive success is calculated based on the same year's spring population size, not for example the following year's size. (The autumn population size, by the way, could also be calculated by taking the total spring population size, $n_n(t) + n_m(t)$, and multiplying it by $1 + a - b(n_n(t) + n_m(t))/2$. Why is this?)

When we now remember that a fraction x of autumn birds go and migrate, and probabilities to survive a winter are s_m and s_n for migrants and nonmigrants, respectively, we can now deduce the *next* year's population sizes, denoted $n_n(t + 1)$ and $n_m(t + 1)$:

$$n_m(t + 1) = [n_n(t) + n_m(t) + n_n(t)R_n(t) + n_m(t)R_m(t)]xs_m \qquad (7.3a)$$

$$n_n(t + 1) = [n_n(t) + n_m(t) + n_n(t)R_n(t) + n_m(t)R_m(t)](1 - x)s_n \qquad (7.3b)$$

Now comes the clever part: if the population is at equilibrium, it will neither grow nor shrink, and the numbers of different types of individual (survived migrants, or survived nonmigrants) stay constant too. This means that $n_m(t + 1) = n_m(t)$, and $n_n(t + 1) = n_n(t)$. In other words, we can forget everything about counting which year things happened in, because we are now interested in a stable scenario where every year is like the previous one. Equations (7.3a,b) now take the form

$$n_m = [n_n(1 + R_n) + n_m(1 + R_m)]xs_m \qquad (7.4a)$$

$$n_n = [n_n(1 + R_n) + n_m(1 + R_m)] (1 - x)s_n \qquad (7.4b)$$

Solving for the equilibrium sizes can now be done, if we assume that the resident strategy x and the characteristics of the habitat, a and b, are known.

Why do we 'know' x? Because we are interested in the consequences of a specific resident strategy. These consequences are the environment that the resident strategy creates for a mutant who tries to invade. Since we will examine every possible resident strategy for invasion prospects of every possible mutant, we will set up each checkpoint by picking a particular value of x, and examining the consequences. In other words, x is known only in the sense that we promise to look at every possible value of x, within the limits of our chosen numerical accuracy (exactly as we did for possible plant heights in the preceding chapter).

We also already know that $R_n = a - bn_n/2$ and that $R_m = a - bn_n - bn_m/2$. There remains, unfortunately, a little bit of algebra to be done: substituting these values for R_n and R_m in Eq. (7.4), solving for n_n and n_m, and simplifying – to find the solutions for n_n and n_m. In practice, people nowadays tend to use programs such as MATHEMATICA or MAPLE to give solutions, but even then a little bit of rearranging is often needed before the most concise form of the solution is found. Consequently, the pen and paper method works almost equally well for a not too freakish set of equations, such as Eqs. (7.4a,b). Either way, we get

$$n_m = 2s_m x \frac{(1 + a)(s_m x + s_n(1 - x)) - 1}{b(s_m x + s_n(1 - x))^2} \qquad (7.5a)$$

and

$$n_n = 2s_n(1 - x) \frac{(1 + a)(s_m x + s_n(1 - x)) - 1}{b(s_m x + s_n(1 - x))^2} \qquad (7.5b)$$

A less interesting alternative, $n_n = n_m = 0$, also solves Eqs. (7.4a,b) and is, therefore, an equilibrium: extinct now, extinct forever. We can leave this aside as this equilibrium offers little insight beyond triviality. This is captured in mathematical jargon too: $n_n = n_m = 0$ is indeed called a *trivial equilibrium*.

It is worth pausing for a while to marvel at the achievements so far. From Eqs. (7.5a,b) we see that the expressions for the spring population sizes n_m and n_n share similar components: for example, both decrease if b

increases. This is quite intuitive when we remember that large b meant a steeper decline in Fig. 7.2, in other words large values of b diminish the number of available territories and make them poorer as breeding habitats. Also, the terms in Eqs. (7.5a,b) have been neatly arranged so that similar components can be compared. The only differences between the expressions for n_m and n_n are the terms $s_m x$ and $s_n(1 - x)$ before the complicated mess that follows. Therefore, a large fraction of individuals making the migration decision (x), and high survival of migrating individuals (s_m), both increase the number of migrants in the spring compared with overwintered birds; high s_n and low x have the opposite effect of increasing numbers of nonmigrants present in the spring. Again, this is very intuitive.

What about the effect of a? We can see that we cannot get a stable population where $n_m > 0$ and $n_n > 0$, unless the numerator $(1 + a)(s_m x + s_n(1 - x)) - 1$ is positive. In other words, $(1 + a)(s_m x + s_n(1 - x))$ must exceed the value 1. This makes very good sense: $1 + a$ is the average number of individuals (parent plus offspring[3]) that emerge from the best possible territory in the autumn, and $(s_m x + s_n(1 - x))$ is the expected survival of a randomly picked individual. If the best territories (Fig. 7.2) fail to provide reproductive success that allows, on average, at least one individual – parent or offspring – to survive to the next breeding season, we cannot have a persisting population at all.

Now that we know n_m and n_n, we have also found out the values for R_m and R_n: remember that $R_n = a - bn_n/2$ and $R_m = a - bn_n - bn_m/2$. Now, we can finally proceed to investigating whether the fraction x, which is the resident strategy in the population, is stable, or if more migration or less migration is selected for. If we have performed our calculations correctly, Eq. (7.1) will give the value 1 for the resident strategy: populations

---

[3] In all these calculations, we operate within a single sex and essentially count reproductive success as female–female reproduction. The reason why we include same-sex offspring only is that including all offspring would necessitate halving their genetic value to the parent when calculating fitness, owing to diploid reproduction. Double the offspring, but each half as valuable – that amounts to the same as same-sex offspring, valued equally highly as the parent, as long as we're assuming a 1:1 primary sex ratio, which we implicitly do here. What about calculations of population stability? This requires that the numbers of each sex remain constant, so considering the dynamics of a single sex works well there too. Can this simplification be always justified? No. In questions where sex ratios themselves are under selection, in cases with weird genetics (e.g. haplodiploid systems) or where adult sex ratios have an effect on density-dependence because, say, males consume more resources, the dynamics of the two sexes will have to be modelled explicitly. For an introductory example see Mangel (2006, pp. 10–12).

remain stable, after all.[4] A mutant, by comparison, will contribute a different number of individuals to next year's population: mutant fitness in one year can be written as

$$w(x') = x's_m(R_m + 1) + (1 - x')s_n(R_n + 1) \qquad (7.6)$$

The important assumption about mutant invasion, and one that makes calculations much simpler too, is that mutants are initially rare. The jargon of the 'rare mutant' means that the value of R_m is calculated using the resident strategy x, rather than the mutant strategy x', even if we are busy calculating the fitness of the *mutant*. Why? Because we are asking whether mutants can invade if they are introduced in small numbers. Being rare, they do not significantly alter the total numbers of competing individuals; they simply experience the environment (which is reflected in R_m and R_n) as set by the resident strategy x. For example, we can ask if invasion is possible with a 30% propensity to migrate, $x' = 0.3$, if currently a 50% migration propensity is the rule ($x = 0.5$). Of course, this 30% rule, if it invades, can then become the new resident strategy, and then we can ask if further reductions in the tendency to migrate are favoured: for example, $x' = 0.25$ when $x = 0.3$, and so on, for every possible pair of residents and mutants. The result can be summarized in a PIP (Box 7.1). If a mutant is able to invade (its fitness is greater than the resident's fitness), we mark the point $\{x,x'\}$ with a '+', otherwise with a ' − '. For clarity, the regions that share the ' − ' sign are usually coloured red, or in black-and-white figures with a dark hue, to indicate a 'no-go zone'.

---

[4] Strictly speaking, we should also check that the equilibrium is not an unstable one, similar to that encountered in Ch. 2. when dealing with the absence of an allele that is selected for. The system will not return to an unstable equilibrium (e.g. absence of allele) if it is slightly perturbed from it (e.g. beneficial allele is introduced). In the case of territorial birds, an example of an unstable population is found if removing only one bird – only once – from an equilibrium population marks the beginning of a permanent decline of the population (without any other adverse environmental changes). This is not impossible, but it requires an Allee effect, which means that *per capita* growth is reduced at low densities (Courchamp *et al.* 1999; Stephens and Sutherland 1999). This is not the case in our example, where *per capita* reproductive success improves at low densities (Fig. 7.2). But since there are also other processes that can lead to instability, such as extremely high fecundity, it is good to know how to look at the stability of equilibria (e.g. Chs. 5, 7 and 8 of Otto and Day (2006) give precise instructions). In fact, our trivial equilibrium $n_n = n_m = 0$ is unstable: adding birds to a currently extinct population would allow the population to bounce back and thus move away from the trivial equilibrium. Of course, this happy conclusion of our model ignores demographic stochasticity, inbreeding and other problems that real reintroduction programs have to take into account. Clearly, we would model differently if we were interested in the details of population dynamics close to extinction.

Box 7.1

How to solve the migration problem and to draw PIPs in MATLAB. The function `migrate.m` produces Fig. 7.6a with the command `migrate([.5 .42],2,0.0002)`. It uses the function `pip.m` to draw the PIP, which is also given below. Both functions should be saved using their respective names before `migrate.m` is run.

```
function [fit_migr,fit_nonmigr,nm,nn]=migrate     } one
(s,a,b)                                           } line
sm=s(1);sn=s(2);
x=linspace(0,1,101);
nm=2/b*sm*x.*((1+a).*(sm*x+sn*(1-x))-1)./...
   (sm*x+sn*(1-x)).^2;
nn=2/b*sn*(1-x).*((1+a).*(sm*x+sn*(1-x))-1)./...
   (sm*x+sn*(1-x)).^2;
% check if some extra modelling work is needed before
% these parameter values can be used
if any(a-b*(nn+nm)<0)
   disp('Warning: some individuals cannot breed!');
   disp('Better not to trust the program output right  } one
      now.');                                          } line
end
% calculate average reproductive success: first
% nonmigrants, then migrants
Rn=a-b*nn/2;
Rm=a-b*nn-b*nm/2;
% fitness of nonmigrants and migrants in one year
fit_nonmigr=sn*(Rn+1);
fit_migr=sm.*(Rm+1);
% then plot everything of interest
figure(1);
subplot(2,1,1);plot(x,nn,x,nm,x,nm+nn);axis([0 1  } one
   0 10000]);ylabel('Spring population size');grid  } line
subplot(2,1,2);plot(x,fit_nonmigr,x,fit_migr,
   'r');ylabel('Fitness');xlabel('Proportion of      } one
   migrants');grid                                   } line
figure(2);
xmut=linspace(0,1,101);
```

Box 7.1 cont.

```
for i=1:101,
  for j=1:101,
    mutfit(i,j)=xmut(j)*fit_migr(i) + (1-xmut(j))*
      fit_nonmigr(i);
    resfit(i,j)=x(i)*fit_migr(i)+(1-x(i))*
      fit_nonmigr(i);
  end;
end;
pip(x,xmut,mutfit-resfit);
```

one line

one line

The function `pip.m` given below can be used in two ways – either in a colourful version that marks 'cannot invade' as red, 'can invade' as green, and 'no difference' as white (note that the diagonal in a PIP should be white); or in a version with shades of grey, which is much better for colourblind researchers. The latter version is brought into use by calling `pip(x,xmut,mutfit-resfit,1)` instead of `pip(x,xmut,mutfit-resfit)`. This sets colourblind to have the value 1, which is interpreted as 'true' in logical comparisons (see Lesson 4 in the Appendix). In Fig. 7.6, the colours have been replaced by '+' and '−' regions, but the message remains the same.

```
function pip(x,y,invfitness,colourblind)
% function pip(x,y,invfitness,colourblind)
% Draws a PIP in Matlab. Gives the range of x and y values
% as vectors, and invfitness as a matrix. The fourth
% argument, which is optional, creates a greyscale
% plot which is easier to interpret for colourblind
% people.
% colours are 1,2,3 for negative, zero, positive sign
% of invasion fitness
z=sign(invfitness)+2;
% if fitness is not a proper number, mark it as colour 4
% which is grey
z(isnan(invfitness))=4;
% the following colour definition uses red-green-blue
% values to mark negative, zero, positive, and NaN
% (not-a-number) as red, white, green, and grey,
% respectively
colour=[1 0 0; 1 1 1; 0 1 0; 0.5 0.5 0.5]
```

```
% Note: 'nargin' means the number of arguments that
% have been given as input; if this is 4 and the variable
% colourblind is true then we'll redefine colours
if nargin==4 & colourblind
   colour=[.8 .8 .8; 1 1 1;
           0 0 0; .5 .5 .5];
end;
sx=length(x); sy=length(y);
% just in case the matrix is arranged the wrong way,
% swap it
if sx~=size(z,1) z=z';
end;
% define the size of the boxes to be drawn
dx=(x(2)-x(1))/2; dy=(y(2)-y(1))/2;
% clear the figure, define the range of values where
% plotting will happen, and tell Matlab to hold on i.e.
% not redraw the whole figure when one box is added
clf; axis([x(1)-dx  x(sx)+dx  y(1)-dy  y(sy)+dy]);
   hold on;
   for i=1:length(x),
      for j=1:length(y),
      xcoord=[x(i)-dx  x(i)-dx   x(i)+dx  x(i)+dx]
      ycoord=[y(j)-dy  y(j)+dy   y(j)+dy  y(j)-dy]
      fill(xcoord,ycoord,colour(z(i,j),:));
   end;
end;
hold off;
```

An example is shown in Fig. 7.5. The interpretation? If the resident strategy is one with $<35\%$ migration, then mutants are able to invade if they migrate more than what is the current population-wide rule (the '$+$' regions lie above the diagonal, i.e. where $x' > x$). But in populations where 35% migrate, mutants are able to invade if they migrate less often than what is the population-wide rule. To summarize, selection favours more migration if 'too few' individuals migrate, and less migration if 'too many' migrate. Hurrah! We have established frequency-dependent selection that can explain why partial migration with 35% migrants can evolve.

For those who take delight in staring at equations: note that in Eq. (7.6) fitness depends linearly on x'. In other words, '$+$' regions

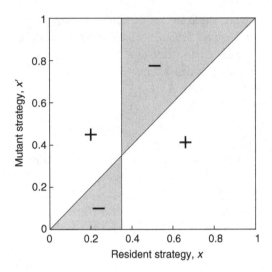

Fig. 7.5 Pairwise invasibility plot with $s_m = 0.6$, $s_n = 0.5$, $a = 1.2$ and $b = 0.0002$. Shaded areas indicate {resident, mutant} comparisons in which the mutant cannot invade; white areas indicate that the mutant can invade.

prevail wherever $x' > x$ and $s_m(R_m + 1) > s_n(R_n + 1)$, or wherever $x' < x$ and $s_m(R_m + 1) < s_n(R_n + 1)$. More migration is selected for when the annual success is greater when migrating, $s_m(R_m + 1)$, than when overwintering, $s_n(R_n + 1)$. The frequency-dependence is created through the fact that R_m and R_n are not constant but respond to x.

In Fig. 7.5, the survival of nonmigrants ($s_n = 0.5$) fell clearly below that of migrants ($s_m = 0.6$), yet an overwintering attempt paid off often enough that we can predict that some individuals should attempt overwintering. To extract the maximal biological insight, it is useful to have a look at some additional examples and plot the fitness associated with the migratory and nonmigratory options, as well as the associated PIP. Let's assume that migration is successful with 50% probability: $s_m = 0.5$. As to the habitat, let us say $a = 2$ and $b = 0.0002$, which means that there are 10 000 territories (since $2 - 0.0002 \times 10\,000 = 0$). The values of n_m and n_n can then be calculated from Eqs. (7.5a,b), and the fitness of strategies follows from Eq. (7.6). Figure 7.6 summarizes the results for four different values of nonmigrant survival, s_n. In all of the examples, we assume $s_n < s_m$ because these are the only interesting values. Why? If s_n exceeds s_m, nonmigrants survive better as well as gain access to better territories. It is then obvious that no bird should ever migrate under those circumstances, and modelling is hardly needed to state this conclusion.

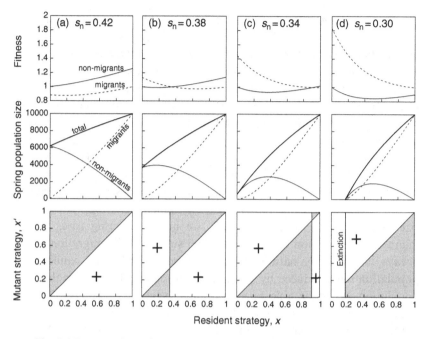

Fig. 7.6 Top row shows fitness in one year for migrants, $s_m(R_m + 1)$, indicated by dotted lines, and for nonmigrants, $s_n(R_n + 1)$, solid lines, for different values of nonmigrant survival: (a) $s_n = 0.42$, (b) $s_n = 0.38$, (c) $s_n = 0.34$ and (d) $s_n = 0.30$. In all cases, $s_m = 0.5$, $a = 2$, $b = 0.0002$. Middle row shows solutions for n_m (dotted line), n_n (solid line) and the total spring population size $n_m + n_n$ (thick solid line). Lowest row shows pairwise invasibility plots indicating partial migration where vertical lines are present. Shaded and white areas indicate mutant invasion prospects into a given resident population, as in Fig. 7.5.

In Fig. 7.6a, nonmigrant survival is 16% lower than that of migrants ($s_n = 0.42$, $s_m = 0.5$). Yet, regardless of the value of x, lower values of x (i.e. less migration) are always selected for: this is seen in that nonmigrant fitness always exceeds migrant fitness, and in the PIP, where the '+' region is the area below the diagonal. In other words, any mutation that makes individuals less likely to migrate than the currently persisting rule in the population is favoured by natural selection. Our conclusion from this graph is that the prior residence effect can be quite a powerful force maintaining overwintering populations. Only once the survival of non-migrants is much lower than that of migrants (Fig. 7.6b, 24% lower; Fig. 7.6c, 32% lower) do we find partial migration evolving. The position of the vertical line in the PIP indicates the point where the migratory and the nonmigratory option yield equal fitness, and there is no selection to increase or decrease migratory tendencies. In Fig. 7.6b, this happens at roughly 30% migration, and in Fig. 7.6c, roughly 90% of birds should migrate. Finally, in Fig. 7.6d, overwintering attempts are so rarely successful that the population cannot persist at all if there are not at least 17% migrants present. It is no surprise, then, that the population evolves towards the migratory option (PIP: the '+' region is above the diagonal). It can be shown that this conclusion is quite general: when the prior residence effect is operating, complete migration can only evolve if a nonmigrant population cannot persist (i.e. nonmigrants would form a sink population; for details see Kokko and Lundberg (2001)).

It is also quite interesting to have a look at the equilibrium population sizes created by each resident strategy. In all cases described in Fig. 7.6, the more migrants in the population, the higher the overall population size. This is quite logical: after all, we are considering cases where migration is the option that confers higher survival. Still, migration is not always favoured by natural selection. Consequently, the equilibrium population size produced by selection often remains lower than that the population would be able to achieve if it behaved differently. This is reminiscent of the conclusion in Ch. 6: solutions of games are often not optimizing the greater good, or are not 'good for the species'. Selfish behaviour is favoured, which in the present case means that the intraspecific competitive rush to occupy best territories makes individuals invest in traits that exist simply to outcompete other conspecifics. In games, the fitness of individuals is generally not maximized in the same simple way as in optimization approaches (Ch. 4).

Finally, let us have another look at our assertion that population dynamics play a role in determining which strategies should evolve. Take Fig. 7.6b, where we predict $x^* = 0.38$ when the best territories have reproductive success $a = 2$. Repeating all the calculations for $a = 1.5$ leads to a case where migration is always selected for; in other words, $s_m(R_m + 1) \geq s_n(R_n + 1)$ regardless of the value of x (the PIP will look identical to that in Fig. 7.6d). Therefore, reducing habitat quality from $a = 2$ to $a = 1.5$ made all birds migratory and destroyed the stability of partial migration. Why? Remember that overwintering was the more risky option, and one could argue that it is no longer worth risking one's life for a territory that offers, at best, only 1.5 offspring per season. But there is more to it: reducing habitat quality means that equilibrium population sizes diminish, and when there is less competition for best habitats, migrants too can enjoy almost as good reproductive success as nonmigrants. The eco-evolutionary feedback (Fig. 7.1) is not just a mathematical construction to ensure that models remain self-consistent – it really does produce predictions where we expect evolutionary changes in the expected behaviour of individuals depending on environmental conditions (Ferrière *et al.* 2004a).

Fitness in one single year only: a caveat?

Before we move on, it is perhaps wise to address a doubt that might be nagging an alert reader. What to do if you are not convinced that improving one's behaviour in 1 year is equivalent to overall improvement of fitness? After all, in Ch. 5 we saw that short-term fitness improvements do not necessarily lead to the best lifetime performance. However, there we were interested in different states (lean, fat) that the individual could end up in as a result of current actions. But in our current case, we ignore such complications. Instead, after 1 year of survived life the bird is again in exactly the same state as before: it is simply a bird who is alive and has just completed a breeding attempt, and it carries no life-history baggage from last actions. This means that it is ready to perform the same improvement over and over again, which should convince you of the equivalence that 1-year improvements lead to overall improvements too.

But if this verbal argument is not enough, the general guidance on modelling philosophy (Ch. 1) applies here too: doubt everything, until fully convinced, and use maths to check things out if necessary. Consider the following calculation, which is not specific to our migratory example, but a little bit more general: a mutant is using a strategy that leads to annual survival s' and reproduction R', and it tries to invade a resident population where a different strategy is employed, such that survival and reproduction equal s and R for residents. This can be thought to reflect a general life history trade-off between s and R, if reproduction is a decreasing function of survival (as it is in our migration example: migrants survive better but cannot reproduce in the most profitable habitats). The resident produces, on average, $s(R+1)$ individuals over the course of a year: it first survives with probability s, and if it survived, it contributes R offspring plus itself to the next year's population. At the ecological equilibrium, population numbers do not change from one year to the next, hence $s(R+1)=1$. We would like to prove that the condition that mutants are superior to residents in one year, $s'(R'+1) > s(R+1)$, is equivalent to an improvement of the whole LRS, $s'R'/(1-s') > sR/(1-s)$. Well, since $s(R+1)=1$, the first condition becomes $s'(R'+1) > 1$, and the second $s'R'/(1-s') > sR/(1-s) = s(1/s-1)/(1-s) = 1$. Therefore, we're testing if $s'(R'+1) > 1$ is equivalent to $s'R'/(1-s') > 1$. This is true, since both conditions can be written in the identical form $s'R' > 1 - s'$, as long as we focus on biologically feasible survival values $0 < s' < 1$.

But does this also hold true for our example where individuals can flip-flop between two different survival values, which necessarily creates some additional stochasticity in lifespan? Well, if not convinced ... again, you must calculate it to convince yourself. In general, the level to which things are proved depends on what one regards as obvious enough. Can one be confident that there are no mistakes lurking, even if every step of reasoning is not spelled out to the greatest detail? There is no unique standard answer as to what level of obviousness one should reach. Mathematics, like any science, is a structure where advanced results rely on more basic theory. The unfortunate consequence is that beginners may find it hard to follow many texts. The only remedy is to work through many examples until the patterns start to feel familiar. It eventually becomes much more easy to jump from one line to the next. But, because such confidence only comes with experience (which, unfortunately, requires some work), one would perhaps want more detail spelt out in published work. Still, going all the way to the other extreme would be unhelpful too. I know this, because during my own mathematical training I had to go through a formal proof that $1 = 1$. This equality is not part of the axioms that define the behaviour of numbers, so one has to make sure that it follows from the more fundamental axioms. The calculation of $1 = 1$ is not too tedious, but still it takes a while to complete. Certainly no evolutionary or ecological journal would like to see that level of proof in any appendix they are willing to publish! So, one has to try to make things 'clear enough', and the definition of 'enough' will depend on the intended audience.

7.2 Got interested?

The idea that partial migration is not necessarily hedging one's bets against variable environmental conditions, but can result from intra-specific competition under density-dependence, was first outlined by Kaitala *et al.* (1993). They also considered many factors not included in the present model, such as possible age dependency of migratory decisions. For an alternative formulation of partial migration in a self-consistent framework, see Kokko and Lundberg (2001). The importance of density-dependence as an important determinant of evolutionary processes is emphasized by numerous authors in various contexts. The following lists only some of a few starting points and illustrative examples: Eadie and Fryxell (1992), Metz *et al.* (1992), Clark and Yoshimura (1993a), Mylius and Diekmann (1995), Kokko and Sutherland (1998), Pen (2000), Pen and Weissing (2000), Claessen *et al.* (2004), Härdling and Kaitala (2005), Kokko and Rankin (2006) and Dieckmann and Metz (2006). For a discussion of self-consistency in modelling (e.g. in the context of parental care, but also otherwise), see Houston *et al.* (2005) and Houston and McNamara (2005). The latter of these papers also discusses Maynard Smith's original model (Maynard Smith 1977). Houston and McNamara (1999) provided a clear discussion of more advanced game theory models in general, including those in which the state of an individual can change over time. Several chapters in Otto and Day (2006) deal with various aspects of forming and solving conditions for equilibria such as our Eqs. (7.4a,b), and their Ch. 12 presents additional examples of invasion analysis. Other useful starting points are Mangel (2006, e.g. Chs. 4–6), and Bulmer (1994, Ch. 5).

The PIPs produced here always had vertical lines whenever partial migration was found. Not all PIPs share this feature: for example, the partial migration model presented here will no longer have exactly vertical lines if the decision to migrate depends on an individual's condition; still, the '+' and ' − ' regions are interpreted as above, and the biological conclusions remain very similar (a keen reader might want to try this out, although the modifications required are quite substantial). The shapes of the ' + ' and ' − ' regions can be diverse and inspecting them yields useful information on various stability properties of the evolutionary equilibria. For a clear classification of the consequences of different shapes, and a discussion of different concepts of stability involved, see Dieckmann (1997), Rousset (2004, Ch. 5) or Otto and Day (2006, Ch. 12). Meszéna *et al.* (2001) have provided beautiful (although at first sight perhaps scary) graphical illustrations of the concepts that lead to a PIP.

The relationship between adaptive dynamics and other modelling methods is a hotly debated topic. A whole issue of the *Journal of Evolutionary Biology* was devoted to this debate in 2005, with Waxman and Gavrilets (2005) providing the target review that many others responded to. Among the articles published in that issue, a must-read contribution – and one that extends beyond adaptive dynamics per se – is that of Butlin and Tregenza (2005). They raised several important points, such as the publication pressure that forces theoreticians to show that 'interesting' outcomes are possible in the same vein as empiricists hope for low enough *P* values to give support for an exciting new theory. They additionally discuss the difficulties of communication between theoreticians and empiricists. The latter often do not possess sufficient knowledge of all the assumptions underlying a particular modelling method. If the reader has to rely on verbal explanations and particular output graphs chosen by the author of a modelling paper, the message of a paper may become over-simplified to the point that scientific progress is led astray. (If reading this book helps at least one empiricist in interpreting at least one important theoretical paper more thoroughly than he or she would otherwise have achieved, I will consider my effort worthwhile. Let me know.)

8

Individual-based simulations

*where virtual butterflies try to fly out of our reach, until
ruthless exploitation of student labour finally captures them*

Perhaps you are one of the readers of this book with a specific question in
mind: 'looks interesting . . . but does it help to answer how I should model
my own problem concerning my favourite snail/bird/mite/ungulate/
whatnot'? And perhaps you have already seen the light and decided that a
population genetic approach will work brilliantly, or perhaps dynamic
optimization, or some other technique introduced so far. But perhaps
I am being overoptimistic here. Quite likely, there are still too many
factors floating around in your head: the organism may be beautiful to
work with, but to model, oh no . . . it does not only X but also Y and Z,
and F and G and K surely influence what it should be doing, besides, the
weather keeps changing and this drastically changes reproductive success
from year to year, so how can one ever hope to summarize the long-term
fitness prospects in a neat simple equation . . .

Well, there are two answers. One is simply to read Ch. 1 again and
think about what is the essence of your problem. Maybe you should start
off by examining one question at a time, rather than trying to answer
everything simultaneously. But, there is also answer number two: if you
still cannot write down an equation despite being clear-headed about
what the question is, and you not being too greedy about how many
factors to allow to influence fitness . . . then you may want to resort to
simulations.

If you feel like asking 'haven't we been simulating things for the last
so many pages already', the answer is, not really. We have been looking
for either analytical or numerical answers to questions, but the answers
have always been exact: for example, on p. 158 we produced an answer

$x^* = 0.38$, which is not an average of many trials but (within the range of numerical accuracy) The Absolute Truth that follows directly from our assumptions. In other words, every single run of the numerical procedure will always produce $x^* = 0.38$. By contrast, simulations usually refer to calculations where random events are allowed to influence solutions, and each run of the simulation, therefore, produces a slightly different answer.

For example, consider the question, what is the probability that an individual reaches the grand old age of 10 years, when annual mortality is 50%. It is certainly possible to give a numerical, exact answer to this: it is $0.5 \times 0.5 \times 0.5 \times 0.5 \times 0.5 \times 0.5 \times 0.5 \times 0.5 \times 0.5 \times 0.5 = 0.5^{10} = 0.00097656$. It is even possible to give an analytical answer (i.e. one that generalizes to all possible values): denoting mortality by d, and the number of years by t, survival after t years must equal $(1 - d)^t$. Analytical results are nice, because they can then be employed in more complicated statements, and results will be trustworthy in a beautifully transparent way. But what if finding the general solution $(1 - d)^t$ was beyond your powers of imagination? You could take a coin, throw it 10 times, and see if you got heads (to mark survival) every single time. If this happened, mark the trial as a 'success'. If it did not, mark it as a 'failure'. Then, repeat this 10 000 times, and you will be quite likely to find around 10 trials in which all 10 throws are a success (for the reason, see the law of large numbers on p. 39). This gives you an estimate of survival over 10 years, $10/10\,000 = 0.001$.

This is what is meant by a simulation. A simulation specifies the rules that govern the lives of individuals and then follows their lives when they try to deal with these rules. Random events typically play a role here, and the term 'Monte Carlo simulation' is often used when one wants to make it explicit that the simulation uses random numbers and, therefore, unfolds in a different way each time it is run. The term 'individual-based simulation' is used when individual fates are tracked in the model rather than, say, average lifespan only.

The above coin-flipping experiment illustrates two points. Firstly, it is not true that one has to resort to simulations whenever chance events are included in the model (random death, weather fluctuations, and so on). It was still possible to get an exact number for the probability of survival, $(1 - d)^t$, without any of the boring coin-flipping. All the models so far in this book have been *deterministic*; in Ch. 7, for example, we made the assumption that offspring production can be precisely predicted based on the spring population size and the qualities of available territories.

Random fluctuations were considered irrelevant to the question, and it is indeed conceptually important to show that evolution could produce a strategy with some random elements – some birds migrate, others do not, apparently at whim – without any randomness of the environment. However, it is often perfectly possible to consider *stochastic* models (i.e. ones that include random variables) and still be able to derive analytical expressions for the likely outcomes. Then, we are not just interested in the mean outcome but also in the probabilistic distribution of possible values. For example, if 100 individuals all have survival prospects $(1 - d)^t$ over a timespan of t years, it is possible to derive the precise distribution of survived individuals: all survive with probability $[(1 - d)^t]^{100}$, 99 survive with probability $99((1 - d)^t)^{99} (1 - (1 - d)^t)$, and so on, using properties of the binomial distribution.[1] And even if the environment varies from year to year, calculations may remain analytically tractable, so simulations are not the only tool.

Secondly, flipping a coin 10 000 times is really boring. Of course, such tasks are all performed by computers nowadays, but still there is some uncertainty left after that: we might get an estimate of 10/10 000 for survival, but 9/10 000 is quite likely too; after much more computing effort, the answer is never as accurate as with numerical analysis or proper analytical equations. Also, the 10/10 000 is a single number; it does not yet give a clue of by how much the number of surviving individuals would have fallen after one more year, whereas the expression $(1 - d)^t$ gives this immediately: change t to $t + 1$ and you have it.

Why bother ever simulating, then? Just to get the survival estimate over 10 years, one indeed shouldn't bother. Rather, training one's mathematical brain sufficiently until one is able to derive answers such as $(1 - d)^t$ leads to much more pleasing results. The reason simulations are useful is simple: often the exact answers are indeed difficult to derive. The biological question can be interesting and yet all attempts to produce analytical solutions fail.... Then, why not just let the computer work out what happens in a virtual world where individuals live, reproduce and die, and try to understand the result.

Two further advantages of simulations also favour their use. One is that, clever as we humans might be, we are still prone to erroneous thought (Piattelli-Palmarini 1994). Therefore, the more complicated our analytical derivation, the more it makes sense to check our results with a

---

[1] This requires that each individual's fate is independent of what happened to the others. For instructions, see Mangel (2006, pp. 88–95).

totally different method: for instance, a tedious (though hopefully computerized) coin-flipping experiment. Secondly, an initial simulation attempt has convinced me more often than once that a particular argument seems to work, and then – after gaining some insight by simulating – I may have hit the correct analytical expression much more easily than I could have initially. To come back once more to the coin-flipping example: if you at first could not see why the survival probability at $t = 10$ equals $(1 - d)^{10}$, simulated trials with different values of d would have helped you to see a pattern.

8.1 Dispersal in a changing world

The Glanville fritillary butterfly (*Melitaea cinxia*) forms continuous populations in some parts of its range. The range, however, also extends to quite challenging environments: the species lives in a network of patches, which are dry meadows dotted around islands that form the Åland archipelago in the Baltic sea between Finland and Sweden. The amount of habitat is small compared with the total area of the landscape, and most patches are so tiny that long-term persistence is impossible in them – sometimes only a single female butterfly might be laying eggs in a patch. The fact that long-term persistence is still possible for the population network as a whole, even if no single patch is guaranteed to maintain a population for long, has spurred a whole branch of population ecology, called metapopulation theory (Hanski 1998, 1999). The key to persistence is that there are always some individuals migrating between the patches, and if this migration rate is high enough, long-term persistence becomes possible.

Note, by the way, that 'migration' means here something different from the go-and-return type of movement typical of bird migration, which was discussed in Ch. 7. For entomologists, migration is often synonymous with what ornithologists would call either natal or breeding dispersal. Nothing new here: throughout the book we have been battling with variations of terminology and notation. Because a successful modeller should keep being interested in questions posed by many different systems, one must learn to master different terminologies as one goes along. In the current chapter, I will use 'migration' and 'dispersal' synonymously.

From a viewpoint of what is 'good for the species' that forms metapopulations, sufficient migration is essential. But we have learnt repeatedly in the various chapters that evolution does not necessarily produce something that is good for the species. How much migration is expected to evolve in a setting where no patch guarantees permanent fitness prospects? Migration could be risky, especially for a butterfly in an archipelago, where unsuitable habitats are much more abundant than suitable ones. But staying put can be risky too, at least if any site can turn unsuitable for reproduction in any particular year. Genes that make individuals stick to where they currently are will sooner or later be wiped out by natural selection: for a butterfly that only lives for a year, 1 year of total unsuitability of the local patch is sufficient to cause extinction for a lineage that has stayed put. And, if there is a nonzero probability that a patch is unsuitable in any given year, this will eventually happen for any patch, so lineages that are too site-tenacious are doomed. There are also other factors influencing the benefits of migration: for example, migrating individuals leave more resources for their relatives who stay put, thus dispersal can be kin-selected (Hamilton and May 1977; Gandon and Michalakis 1999). Additionally, dispersers may also be able to find better habitats, or empty habitat patches, where resources are more plentiful than in their potentially crowded birthplace ...

A lot of things to think about! To make matters worse, in a changing world it would be neat to be able to predict population responses to changes in the environment (Ferrière *et al.* 2004a; Kokko and López-Sepulcre 2006). For example, what if some sites get permanently bulldozed, so that it is now much harder for individuals to find new sites? Clearly, this is bad news for the persistence of the metapopulation; but exactly how bad? There is a real risk that butterflies are selected to disperse less when dispersal is more dangerous, but since the metapopulation relies on suffi-cient dispersal, the 'selfish' sedentariness of individuals could cause the extinction of the whole population (Gyllenberg *et al.* 2002; Parvinen 2004).

An idea put forward in this context is one of *evolutionary rescue* (e.g. Ferrière *et al.* 2004b; Holt and Gomulkiewicz 2004). Perhaps anthro-pogenic damage on this planet can, to some extent at least, be compen-sated by adaptive evolutionary responses in the populations. For example, when predicting the responses of bird populations to changing climates, it is important to know if they are able to adjust their breeding dates so that they remain in synchrony with newly shifted peaks in resource availability (Coppack and Both 2002; Visser *et al.* 2004; Nussey *et al.* 2005). If they can, then declines in population numbers can be prevented or at least alleviated: rescue has taken place. In the context of dispersing butterflies, an evolutionary rescue could occur if more dangerous dispersal was partly compensated by evolution towards higher dispersal rates.

But – we just noted that butterflies are probably selected to disperse less, not more, when dispersal becomes more dangerous. Predicting higher dispersal rates seems downright wrong. Still, the idea is not as crazy as it first seems, when we think about population-level feedbacks (Ch. 7). And, no, we do not even need to abandon the idea that selection is shortsighted and acts primarily at the level of the individual. Here is how it works. If dispersal is very easy, lots of migrating butterflies survive the journey, and this means that a butterfly landing in a new patch is quite likely to be surrounded by other butterflies. This creates competition for local resources, diminishing the expected rewards (offspring production) compared with an empty patch. But if migration is a lot riskier, then there are few dispersers that succeed, and the proportion of patches that offer abundant resources consequently increases. So, while the journey is riskier, the rewards are higher *if* one happens to survive it and find a suitable patch.[2]

But can the heightened reward be strong enough to select for higher dispersal rates, given that the journey really is riskier now? The problem is complex: patches can be occupied and unoccupied; some patches in some years do not allow any offspring production; the number of individuals arriving somewhere depends on the numbers that were leaving patches (and on pure luck); the subsequent offspring production will depend on how many individuals arrived and where, ... Even if it is recommended to stop and think first before resorting to simulations, in this case it is quite clear that simulations can really help to see if evolutionary rescue can take place. Heino and Hanski (2001) have indeed done this, and to make the approach particularly valuable, they have parameterized their model with real data from the *M. cinxia* metapopulation. A long-term study on this butterfly (for a summary see, for example, Ehrlich and Hanski (2004)) has produced estimates for all kinds of parameters such as mean diapause survival of larval groups, and the daily probability of laying an egg patch.

And, indeed, Heino and Hanski (2001) predict that evolutionary rescue is a possible phenomenon: Figure 8.1 shows how migration propensity first decreases but then increases with migration mortality. By the way,

---

[2] In case you are wondering, female butterflies can mate first and then start searching for ovipositioning sites, which means that they do not need to arrive at a patch with a male to be able to found a new population.

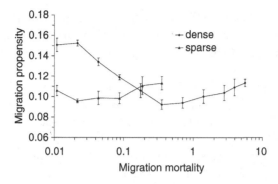

Fig. 8.1 Evolutionary rescue in an earlier model. The *x* axis is a measure of migratory mortality, and the evolutionary response is U-shaped both for dense networks (patches close together) and for sparse ones. (Modified from Heino and Hanski 2001).

do not get shocked by migration mortality exceeding the value 1 in Fig. 8.1. Mortality is not defined as a fixed probability in Heino and Hanski (2001) but as a parameter that influences how likely it is that a butterfly survives migration when starting from patches of varying connectivity – a measure of how close the natal patch is to other patches, how many other patches there are and how large these are.

The model presented by Heino and Hanski (2001) is so rich in detail that it does not make sense to replicate it here faithfully. Instead, let us assume that we get terribly excited about the result and would like to know how easily it generalizes: forget the details about *M. cinxia*, and see if a simpler model yields the same result. Even so, the problem seems to require simulating, rather than resorting to an analytical approach: the list of phenomena (occupied and unoccupied patches, random disasters in some patches, and so on) is long enough that trying to compress it all into simple equations only produces a nasty headache. So, let's simulate, then.

Simulation is about creating a virtual world. What should the world contain? Individuals, and patches, clearly. The individuals must differ in something that we could call an allelic value, if following the jargon from Ch. 3, or strategy (Ch. 6). How to code for this? The simplest choice is certainly to assume that each individual's migration strategy is described by a value 'migration propensity', which we denote by *x*. This equals the probability that an individual moves from its natal patch, whether this is 'intentional' movement or, as sometimes the case for butterflies, simply a consequence of high movement rate that makes it accidentally fly away from a patch.

We could, of course, opt for two values of x per individual – resembling two alleles at the same locus – if we were looking at diploid reproduction. But let's keep everything as simple as possible and forget about this detail (Ch. 2).

What else do we need to know about the individual? Current location is clearly essential, and this could be recorded as an integer number, specifying which patch the individual is in. Other aspects of the life history, such as fecundity, will be tracked by creating new individuals as we go, so we do not need to store that information with our virtual animals: migration propensity and current location is all we need.

So, we are ready to start our population: let's say that there are N sites, numbered from 1 to N, and the initial number of individuals is ... well, how many? Blaming our lack of imagination, let's say this could equal N as well. This is a convenient number because we can start our population with one individual per patch, and this is precisely what we will do: the first one is assigned to patch number 1, the second one to patch number 2, and so on. The individuals will also need to differ in their migration propensity. Before evolution has been initiated in our model, we do not know what values are realistic, or will perform well. We simply need a starting population with different choices for migration propensity. Let's say that each individual has a randomly chosen value of x, uniformly distributed between 0 and 1, as this interval covers the whole range of numbers that can be interpreted as probabilities.

Then, the population is ready to rock'n roll. What happens in one generation of our virtual world? Some sites get destroyed. Individuals move about. At some stage, there must be competition for space. Reproduction must take place. We will also have to think about possible mutation – otherwise our simulation will be endlessly limited to operating with the initial values of x that the random number generator happened to throw onto the stage. We will have to decide in which order it makes sense to perform all these calculations: mutations in the migratory propensity, for example, should happen pretty much immediately once offspring are formed, but many other choices are less fixed. And, at some point, the order in which individuals are allowed to do something (e.g. find a breeding site) should be randomized. This is to avoid weird effects where a strategy ends up artificially more successful than others because individuals who appear first in a list are allowed to choose first.[3] Finally,

---

[3] A related – and worrying – phenomenon was observed by Tregenza (1997): scientists whose last name begins with a letter that is listed early in the alphabet are cited more than

we should perform some data collection each generation: simulations produce so much data that storing it all is usually not feasible. For the sake of nerdy entertainment, some of this data can be plotted 'as we go'.

The above list is quite long: lots of decisions to make before a single evolutionary step has been taken. The advantage of simulations is that a lot of detail can be included without too much effort (because no equations have to be solved beforehand), but the downside follows immediately: unless we have an exceptionally well-parameterized system, many of our choices will have to be made quite arbitrarily. We shall return to this issue later, but let's first have a look at a program that performs everything we asked above. Whether interested in the nitty-gritty details of MATLAB or not, you will immediately notice that the program (Box 8.1) is longer than most presented in this book: the many details have taken their toll.

Box 8.1

The MATLAB program to produce the simulation; this should be saved as `dispersal.m` before running it. For example `dispersal (100,1,0.01,2,0.01,0.1)` produces a result that should resemble Fig. 8.2a; note that a simulation output will be different every time it is run, assuming a sensible random number generator. There are some MATLAB tricks involved to keep the program reasonably short: for example, `unique(pop(:,2))` gives a list of all the sites in which one or more individual resides; the `unique` function is provided by MATLAB and it creates a list of unique numbers that a vector contains. We use this function to ensure that each site is listed only once, and the length of this 'unique' vector therefore gives the number of sites occupied. Likewise, `find` is used to tell the location of elements in a vector for which a particular condition applies, and `intersect` is used to find out identical elements in two different vectors (which can be different in length).

The function also includes an option of suppressing the graphical output. Plotting graphs is time consuming, which is why one does not want this to happen when calculating a large number of results (see Box 8.2).

Note that the function does not prevent that the new location of a dispersing individual could accidentally be the same as the old one.

---

others, perhaps because people often pick the first few suitable references from a list of candidates.

Box 8.1 cont.

Perhaps it could be argued that a mobile butterfly could accidentally end up in the same place, but if you dislike this argument, feel free to change the program.

```
function [xmean,n,xstd,occupied] = dispersal(N,B,    } one
  p,q,b,m,no_plot)                                     } line
% function [xmean,n,xstd,occupied] = dispersal(N,B,  } one
  p,q,b,m,no_plot)                                     } line
% N sites. Migrants get randomly distributed.
% Competition: only some (B, must be an integer) can
% breed.
% Some sites get occasionally destroyed, probability p.
% Mutation probability is q.
% Breeding success is b, this too should be an integer.
% Migration mortality is m.
% The optional 'no_plot', if set to 1, suppresses the
% graphical output.
% Outputs: the time series of the mean migratory
% propensity in the population and its standard
% deviation over time (xmean, xstd),
% the size of the population (n),
% and the number of sites occupied (occupied).

if nargin < 7 noplot = 0; end; % if the number of
% arguments given to this function was less than 7, then
% we assume the user does not want to suppress the
% graphical output

% initialize population: 1st column = strategy,
% 2nd column = location
pop = [rand([N 1]) (1:N)'];

for t = 1:1000
  % data collection
  xmean(t) = mean(pop(:,1));
  xstd(t) = std(pop(:,1));
  n(t) = length(pop(:,1));
  % plot things
  if ~noplot & t/20 == floor(t/20)
    figure(1);
```

```
  subplot(2,1,1); plot(1:t,xmean);
  ylabel('Mean dispersal rate');
  title(['Mortality=' num2str(m)]);
  subplot(2,1,2); plot(1:t,n);
  xlabel('Time');
  ylabel('Population size');
  drawnow;
end;

% randomize order of individuals using 'randperm'
ind=randperm(size(pop,1));
pop=pop(ind,:);

% destruction of some sites. If a random number falls
% below p then that site will be listed in
% 'disasterplaces'.
disaster=find(rand([N 1])<p);
disasterplaces=intersect(pop(:,2),disaster);
% not that disasterplaces have been found,
% individuals residing in them will cease to exist
for i=1:length(disasterplaces)
  f=find(pop(:,2)==disasterplaces(i));
  pop(f,:)=[];
end;

% competition occurs among the remaining individuals
for i=1:N
  potentialbreeders=find(pop(:,2)==i);
  % how many in site i?
  % if length (potentialbreeders)>B
    % too many try to breed here
    % superfluous individuals are removed
    pop(potentialbreeders(B+1:end),:)=[];
  end;
end;

% collect data:
% how many sites are occupied, thus what proportion is
% empty?
occupied(t)=length(unique(pop(:,2)))/N;

% if the population went extinct, we may stop here
```

Box 8.1 cont.

```
if size(pop,1) == 0 break; end;

% reproductive output is collected in 'newpop'
newpop = [];
for i = 1:b, newpop = [newpop; pop]; end;

% mutation occurs among newborns, with probability q
% for each of them
mutate = rand([size(newpop,1) 1]) < q;

% 1st column of newpop stays the same if there is no
% mutation, but if there is, a new random number is
% inserted
newpop(:,1) = newpop(:,1) .* (~mutate) + ...
    rand([size(newpop,1) 1]) .*mutate;
% an individual disperses to a randomly determined
% location, if a random number falls below the
% dispersal gene value (indicated in the 1st
% column of newpop)
move = rand([size(newpop,1) 1]) < newpop(:,1);
newlocation = ceil(N*rand([size(newpop,1) 1]));

% 2nd column of newpop stays the same if there is no
% movement but if there is, the predefined new
% location is inserted
newpop(:,2) = newpop(:,2) .* (~move) + ...
    newlocation.* move;

% mortality for the ones who moved
death = find((rand([size(newpop,1) 1]) < m) .*move);
newpop(death, :) = [];

% the new generation is ready
pop = newpop;
end;
```

For each time step, *t*, the program

- collects data characterizing the state of the current population: mean and standard deviation of current values of *x* present in the population, as well as the number of individuals present

- plots a summary of what has been going on so far, but only if t is divisible by 20 (i.e. every 20 time steps) to avoid wasting too much time
- randomizes the order of individuals present
- destroys some sites temporarily; this is done by removing all individuals from sites for which a uniformly distributed random number happens to fall below a threshold, labelled p
- calculates local competition over space in all sites; in contrast to Heino and Hanski (2001) where competition was at the larval stage, we assume that an integer number (denoted B) of individuals are allowed to breed (see Hamilton and May 1977) while others disappear (i.e. they die)
- performs a check to see how many sites are now empty – also checks if population has gone extinct by now
- makes remaining individuals reproduce, which is done by adding b clones of the parent to a list of individuals that forms the new generation
- mutates each value of x in the new generation, with a low probability denoted by q, to a new randomly chosen value between 0 and 1
- chooses individuals that move by comparing random numbers to their individual strategy values x; for moving individuals, an integer number between 1 and N is chosen as their new location[4]
- imposes an additional mortality risk m on individuals who have moved, and removes those who are unlucky enough when encountering this risk.

Phew! Quite a feat, but what does it tell us?

A couple of trial runs with the program produce encouraging results. Everything seems to run smoothly in a way that makes intuitive sense: for example, increasing the number of individuals (B) that can share a patch increases the equilibrium population size, which also increases the chances that a patch is found occupied (Fig. 8.2). This also appears to select for reduced migration propensities: even though local population densities are likely to be higher when the equilibrium population size is high, the chances of obtaining a breeding position locally are also improved through an increase in B. It is not much use to move to escape this competitive pressure, as the surrounding area is unlikely to contain many empty sites (Fig. 8.2).

---

[4] To be exact, this means that a fraction $1/N$ of 'moving' individuals do not actually move, because the old and new site might coincide by chance. This deflates the values of x slightly, and a more complete version of the program should correct for this (e.g. by preventing choices where the new and the old site coincide). There are other hidden assumptions that one should make explicit, for example that all sites are equally reachable from any other site, which is yet another difference to the more realistic model by Heino and Hanski (2001).

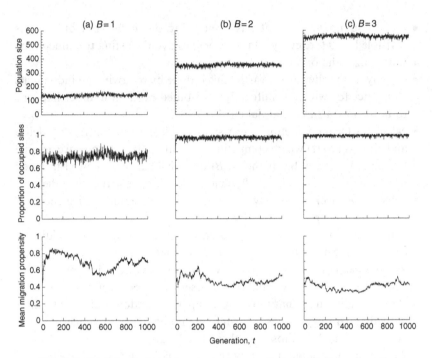

Fig. 8.2 Single-run simulation output regarding population size, the proportion of occupied sites and the mean value of x in each generation. Computations are performed with $B = 1$, 2 and 3, in (a), (b) and (c), respectively. Other parameters: $b = 2$, $p = q = 0.01$, $m = 0.1$ and $N = 100$.

Of course, the wiggly course of evolution in the migration propensities (Fig. 8.2) suggests that a lot of replicates should be run before this result can be confirmed. We started the simulation with a distribution that is probably very far from an ecological equilibrium with its one individual per site, let alone an evolutionary one. For example, had we terminated evolution at $t = 500$, we could have come to a perhaps too rash conclusion that the difference that B makes is quite small; and anyone with the slightest bit of training in statistics would consider the true strength of the effect of B still very much unknown after 1000 years of a single replicate run. One could increase the number of individuals involved by introducing a larger world (larger N). This would help us to make evolutionary trajectories smoother and clearer – but at a cost of more computing time and, for the same amount of computing time, fewer replicates.

But what about our real interest: evolutionary rescue? Evidence for a rescue is found where migration rates evolve towards higher values if the migratory mortality m is increased. So, instead of us varying the value of

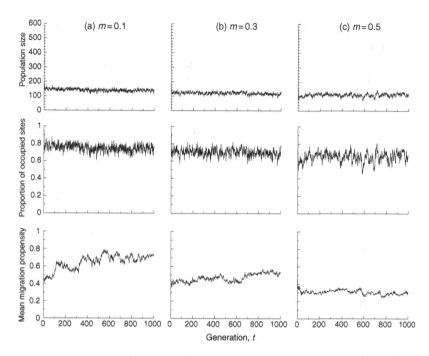

Fig. 8.3 Single-run simulation output, labelled as in Fig. 8.2. Computations are performed with $m = 0.1$, 0.3 and 0.5, in (a), (b) and (c), respectively. Other parameters: $b = 2$, $p = q = 0.01$, $B = 1$ and $N = 100$.

B, let us keep it constant (say, $B = 1$) and, instead, run the simulation with a variety of values for m. Here, our initial trials are not successful (Fig. 8.3): the simulation with a highest value of migratory mortality, m also leads to the lowest migratory propensity, x. The fairly obvious conclusion that selection directly favours less migration when migration is dangerous appears to be stronger than our convoluted argument about site occupancy in a scenario of risky dispersal. But is there any sign that the higher proportion of empty sites, caused by a high migratory mortality, m, should favour migration? There is, indeed, a hint that the proportion of empty sites stays higher when $m = 0.5$ than when $m = 0.1$, although the effect appears decidedly small (Fig. 8.3). The effect simply is not strong enough to counter the fivefold increase in mortality.

What to do to find out if evolutionary rescue can happen in the model? One should clearly do a much more exhaustive search and perform replications with identical parameter values that encounter a different sequence of random numbers each time. To aid our initial search, however, it might be best to do the following: let m take all possible values between 0 and 1; record the mean migration propensity and the proportion of empty sites after single runs at $t = 1000$; and inspect the pattern that forms when plotting these variables against m. This makes our search somewhat more automatic than comparing a couple of haphazard choices for the values of m. With our more systematic approach, we will start seeing signs of evolutionary rescue if we manage to generate a U-shaped plot, similar to that found in Fig. 8.1.

The program that does this 'rescue searching' in a loop trying out different values for the migratory propensity m is called `disploop` (Box 8.2). A first run with $N = 100$, $B = 1$, $p = 0.15$, $q = 0.01$, $b = 2$ (in other words, `disploop(100,1,.015,2,.01)`) produces a plot that looks ... more than a little discouraging (Fig. 8.4A,a). Sure, increasing mortality m does create a marked decrease in the proportion of occupied sites (Fig. 8.4A), which was the requirement that Heino and Hanski (2001) found for an evolutionary rescue to take place. Indeed, any population with $m > 0.5$ could not persist for long at all, so clearly rescue is badly needed here. But any such rescue refuses to happen: below $m = 0.5$, higher mortality simply selects for less migration propensity (Fig. 8.4a), while above $m = 0.5$ it does not make much sense to plot what the population was up to just prior to its extinction, so propensity values for extinct populations have not been included in the graph. To sum up, we have only achieved a slightly more systematic repetition of what we already found in Fig. 8.3.

Box 8.2

How to search for the U-shaped evolutionary rescue. The function `disploop` (saved as `disploop.m`) automatically tries out different values of m between 0 and 1 and summarizes the result of each run as one single dot in Fig. 8.4. Since `disploop.m` takes advantage of `dispersal.m`, both must exist in the same folder of a computer before `disploop` can be run. – Using `disploop` makes the continuous plotting in `dispersal.m` appear rather annoying. This is solved by suppressing the graphical output in `dispersal.m`, achieved by setting the parameter `noplot` to 1 within `disploop`.

```
function [X,E,m] = disploop(N,B,p,q,b)
% function [X,E,m] = disploop(N,B,p,q,b)
% Search for the U-shaped evolutionary rescue, by
% looking at 101 values of m from 0 to 1. Parameters:
% N: number of sites (must be an integer)
```

```
% B: competition, number of individuals per patch who
% can breed.
% Some sites get occasionally destroyed, probability p.
% Mutation probability is q.
% Breeding success is b, this should be an integer too.
m = linspace(0,1,101);

% let's randomize the order in which each value of m is
% tested.
% This is just for entertainment as the plot will
% begin to take shape more quickly than
% when starting from the low end of values for m
m = m(randperm(length(m)));

% Meanx and Occupied will record the outcomes: mean
% dispersal propensity, and proportion of sites
% occupied.
% (Here we use capital letters to remind us that they
% gather the output from a large number of simulation
% runs)
% First make them vectors that are not numbers (NaN = not-
% a-number), but equal m in length, so values that
% are not yet calculated can be 'plotted' (they
% remain empty) along completed ones
Meanx = NaN*m; Occupied = NaN*m;

% then find out what happens ...
for i = 1:length(m)
    i % display which value of i we are calculating now
    noplot = 1; % let's prevent plotting in dispersal.m
    [x,n,xstd,occ] = dispersal(N,B,p,q,b,m(i),          } one
        noplot);                                         } line
    % the last value of the time series is of interest here
    Meanx(i) = x(end);
    Occupied(i) = occ(end);
    figure(2);
    % only plot values of X if there is a population
    % (i.e. Occupied > 0)
    subplot(1,2,1); plot(m(Occupied > 0),               } one
        Meanx(Occupied > 0),'.');                        } line
    axis([0 1 0 1]); xlabel('Migration mortality');
```

Box 8.2 cont.

```
ylabel('Mean dispersal at the end of the              ⎫ one
    simulation');                                     ⎭ line
subplot(1,2,2); plot(m,Occupied,'.');
axis([0 1 0 1]);
xlabel('Migration mortality');
ylabel('Proportion of empty sites');
drawnow; % needed in lengthy calculations to remind
% Matlab to plot now rather than carry on calculating
% more interesting stuff
end;
```

Still, we only chose to examine particular values of N, B, p, q and b. Perhaps the story will be different when alternative choices are examined. Let's pause and think for a moment. For an evolutionary rescue to take place, the reduction in the number of competitors with increasing migratory mortality must play a large role in the life of individuals. Therefore, we could conclude that in any rescue scenario competition must be a serious constraint to begin with; in other words, when m is low, most patches should be occupied so that few vacancies are available. But no value of m produced a particularly high occupancy proportion in Fig. 8.4A, so we should clearly change some of the values N, B, p, q and b to increase this proportion.

Figure 8.4B,b depicts a new trial with $B = 2$ instead of the earlier $B = 1$. The rationale is that a larger number of breeders per site leads to more reproduction and a higher equilibrium population size, thus more competition too. Indeed, the proportion of sites occupied is higher in Fig. 8.4B than in Fig. 8.4A. But does the severity of competition produce evolutionary rescue? The short answer is unfortunately 'no', although stretching one's powers of imagination one could start to see a better U-shape in Fig. 8.4b than in Fig. 8.4a.

What to do next? Can we find examples that are really clear? It is very tempting to increase the value of B further. However, an attempt with $B = 3$ does not yield a U-shape either (Fig. 8.4C,c). At this point, frustration takes over and one tries some other random choice: let's change N to a much smaller value than 100, for example, $N = 20$. The rationale? Perhaps we were wrong about the importance of competition (high occupancy proportion), and perhaps the difficulty of persisting in a small landscape selects for rescue more easily instead. The result: more scatter, but definitely no rescue (Fig. 8.4D,d).

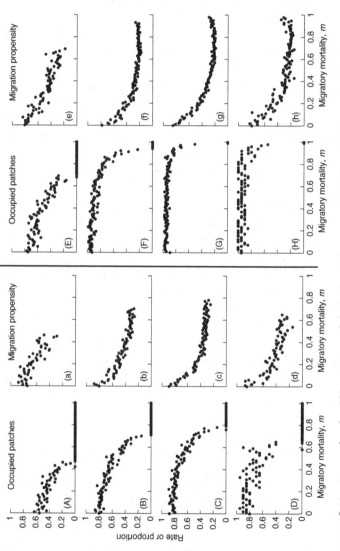

Fig. 8.4 Output of disploop, showing 101 outcomes of single simulation runs, each with a different value of migratory mortality m. Dots indicate the proportion of sites occupied or the mean migration propensity at the final time step of each simulation. Panels labelled with capital letters show the proportion of occupied sites; lower-case letters show the evolved mean migration propensity at the end of the simulation. Values of migration propensity are only shown if a population still persists after 1000 generations. Parameter values: (A–a) $N=100$, $B=1$, $p=0.15$, $q=0.01$, $b=2$; (B–b) $N=100$, $B=2$, $p=0.15$, $q=0.01$, $b=2$; (C–c) $N=100$, $B=3$, $p=0.15$, $q=0.01$, $b=2$; (D–d) $N=20$, $B=3$, $p=0.15$, $q=0.01$, $b=2$; (E–e) $N=100$, $B=1$, $p=0.05$, $q=0.05$, $q=0.01$, $b=2$; (F–f) $N=100$, $B=3$, $p=0.05$, $q=0.01$, $b=2$; (G–g) $N=100$, $B=5$, $p=0.05$, $q=0.01$, $b=2$; (H–h) $N=20$, $B=5$, $p=0.05$, $q=0.01$, $b=5$.

Back to $N = 100$ then, but let's now try something different: reduce p from 0.15 to 0.05, to indicate a smaller probability that a site does not allow reproduction in any given year. The result ... no rescue (Fig. 8.4E,e). At this point, one really is frustrated. As I mentioned in the Preface, this book is the result of a course I have been teaching – and this chapter is the material that I developed for the final week of the course. To cover up my failure to find anything very interesting, I convinced myself that it is perhaps good pedagogic practice that students do not always get exercises that are carefully preselected to produce interesting patterns. (We could thereafter discuss how to react to all those real-life situations where things do not work out as planned.) I gave the material to students, asked them to search for evolutionary rescue and waited to see what happened. I was not surprised to see frowns and shaking heads for quite a while. But within half an hour I also heard enthusiastic cries from the last row, immediately followed by jealous expressions by the others. Yes, the U-shape had been found. And no, I did not know beforehand that it eventually would be.

The secret of success turned out to be sufficiently intense competition created by a combination of high enough B *and* high enough b – in other words, there should be many individuals breeding in a patch, each producing many adult offspring (Fig. 8.4f–h). Whether there are 20 or 100 patches matters very little (Fig. 8.4g,h). It is instead important that vacant patches are very rare when migratory mortality m is low, and that the high level of occupancy is sustained up to fairly high values of m. Because no population can persist when $m = 1$, the above conditions mean that the eventual decline in occupancy with increasing m is steep. This, in turn, means that increasing m leads to a rapid increase in the number of vacant patches in at least some part of the range of possible values of m. Since this response is the basis of evolutionary rescue, it is not difficult to explain why these particular parameter values lead to the rescue.

But this insight is really a hindsight. The dots of Fig. 8.4 still remain examples only, and they should first of all be corroborated by running the same parameter combinations a large number of times, forming confidence intervals for the evolutionary response. Worse, however, the number of different combinations of N, B, p, q and b that remain completely unchecked is necessarily infinite. Figure 8.4 presents eight representative cases, but in reality the students had to check many more before they found the interesting cases. Therefore, were it not for the efforts of eager students, I might have stopped short of finding the particular combinations that led to the rescue. Had this happened, I might have been tempted to conclude that rescue does not evolve that easily: after all,

I really had no guarantee that it would ever occur in this model. But how many combinations (out of the infinity of possibilities) should one have checked before drawing this conclusion and, perhaps, starting to consider the differences between the current model and that of Heino and Hanski to find out what was responsible for the difference? This is tricky: it is always possible that in some dusty corner of parameter values, which we did not yet check, evolutionary rescue indeed operates, a scenario that really turned out to be the case here. Individual-based simulations always share a problem with numerical analyses: as long as one cannot check every single combination of numbers, it is impossible to make statements of the type 'X can never happen', while it is possible to make the more positive statement 'yes, X can happen'.[5] The problem is worse with simulations than with numerical statements, however: checking each corner takes longer, because of the stochastic nature of the outcomes.

Additionally, the number of corners to be checked is usually far greater with simulations than with simpler modelling approaches. Even if our model depicts a simpler world than that used in the butterfly simulation (Heino and Hanski 2001), it remains much more complicated than the models of earlier chapters, and this forced us to make many decisions. One not only chooses the parameter values in each run, but also the rules of the world one is creating. Just to mention one example, we had to specify how much mutant offspring resemble their parents with respect to x. Our choice was 'not at all', since a totally new x was chosen whenever

---

[5] Of course, this problem is not unique to theoretical studies on evolution. It is a tenet of modern science that there is no spontaneous generation of life: maggots do not appear spontaneously in rotting meat, adult flies are required to lay the eggs first. But how to prove this? The first attacks on the idea were made long before Pasteur. Already in 1668, the Italian physician Francesco Redi experimented with sealed and open containers, but the matter was not settled until much later. For example, in 1745 John Needham showed that 'spontaneous' life appeared in chicken broth that had first been boiled and then sealed. An Italian priest, Lazzaro Spallanzani, replied that air could have been contaminating the samples. Sure enough, replicating the experiments by making sure air cannot enter prevented life from appearing. But was this convincing? No, one could argue – and so it was indeed argued – that the experiment only shows that spontaneous generation does not occur without the beneficial effects of air. After Louis Pasteur's extensive experiments in 1859 (the same year that Darwin published his *Origin of Species*) the idea finally died down. Pasteur managed to produce flasks so thin-necked that air could enter but microorganisms could not. But, of course, did these experiments really show that spontaneous generation never occurs? Strictly speaking, no. They only show that life does not spontaneously generate very fast ... at least not at these temperatures ... or not in France. The firmness with which we nowadays state that spontaneous generation does not happen is not based on the outcome of one experiment; instead it reflects our most parsimonious explanation so far, summing up a lot of what we know about life in general. This is not wholly unlike the need to consider the most important questions in evolutionary ecology using different kinds of modelling techniques.

mutation took place. This perhaps does not conform to any particularly elegant genetic model, but then, shouldn't evolution be able to operate regardless of the exact form of mutations as long as there is heritability and some mutation? Yes, it should, and in this sense it is encouraging that both this model and the earlier butterfly simulation show that evolutionary rescue is possible. But had we not found rescue, it would have been difficult indeed to find out what assumption contributed most to the difference between the output of the current model and that of Heino and Hanski. In our model, competition over local resources happened in a different way, the spatial distribution of sites was more abstract than theirs, and so on. If two models show different results and differ in a great number of assumptions, it is obviously very difficult to pinpoint the cause.

The remedy? Whether the model was a 'success' (as happened here) or a 'failure', the next step is once again to stop and think. Did the model give us a view on the most likely conditions under which evolutionary rescue could operate? Here, we have already gained some insight about the role of intense competition and high occupancy rates, but how general is this conclusion? Is there a chance to modify existing models little by little to investigate which factors enable evolutionary rescue to happen, and which ones tend to destroy it? Is there a chance to choose a simpler framework, where we can resort to methods other than messy simulations? One should never underestimate the chances of deriving analytical results even for complicated scenarios.[6] Indeed, an increase in dispersal rate with dispersal mortality has also been shown in an analytical model (Gandon and Michalakis 1999; see also Comins *et al.* 1980) – no mean feat, given the complexity of the question. These results suggest that a more explicit look at kin selection would be worthwhile here, especially since Heino and Hanski (2001) also found that relatedness increased with migration mortality (see also Ch. 7 in Frank 1998).

I have intentionally chosen to present the initial frustrations to highlight the 'cons' side of the 'pros and cons' of simulations. This is for two good reasons. The first point relates to modelling in general. Textbooks often present smooth and beautiful end results only, while 'real' modelling often leads to dead ends and moments of puzzlement, if not true despair, before one finally gains better insight of what is going on. Here,

---

[6] To be fair, it should be said that dispersal evolution is not the easiest example for learning how to gain analytical results. However, it is a very thought-provoking one. If you wish to have a go, the basic starting point is the classic paper by Hamilton and May (1977), and its more extended textbook explanations: Ch. 12 in Otto and Day (2006) and Ch. 6 in McElreath and Boyd (2007).

if our intuition had been that evolutionary rescues should happen easily and a model is just a quick way to prove that to the rest of the world, a dose of disillusionment may turn out quite healthy. An evolutionary rescue is certainly not an off-the-shelf remedy for declining populations, and besides, we did not yet consider how effective the rescue is. This would require more work, comparing the dynamics of populations that are given the chance to evolve migration rates to ones in which these rates are fixed.

The second point applies to simulations perhaps more than to other forms of modelling. Watching their progress can be quite fun, and my experience with students tells that many find it almost addictive. Additional rules can be added *ad libitum*. Programming languages are polite and obedient creatures: they are happy to number crunch without stopping to ask us if doing some extra maths to work out the expected outcome could have been more useful. There is a real risk of getting lost in the virtual world, without noticing that any chances of gaining true insight were lost 2 months of programming effort ago. But as with any good medicines, a healthy but not excessive dose of simulations can do wonders: take it for gaining initial insight into a challenging problem, for verification of analytical results, and for exploring new avenues one would otherwise not have thought about, or been able to model. Numerous times, good ecology has been produced using simulations – but please do read the other eight chapters of this book too.

8.2 Got interested?

Perhaps the most famous simulation exercise in evolutionary ecology is the computer tournament by Robert Axelrod and Bill Hamilton (1981). They were interested in the evolution of cooperation, often modelled as the so-called repeated Prisoner's Dilemma game. This game captures the essence of the difficulty of explaining cooperation, which is that it would be best for an individual if she could let others do all the work (cooperate) but not bother contributing herself (defect), but if everyone realizes this and starts behaving this way, there is no joint effort and nothing is achieved at all. In the Prisoner's Dilemma game, two players meet repeatedly and can either cooperate or defect in each round, and the decision may depend on what they have observed so far. What is the best strategy? Axelrod did not know the answer; one could think that it pays more to cooperate if the opponent is behaving cooperatively as well, but how exactly should one form the decision based on the last *n* encounters? He resorted to an experiment and invited researchers to submit suggestions for strategies that were then competing against each other in a computer tournament.

Fourteen experts submitted entries, some of them describing highly intricate rules of updating one's beliefs regarding the opponent's likely behaviour. But the winner, submitted by Anatol Rapoport, was the simplest of all strategies: 'Tit for tat', which cooperates in the first round and forever afterwards copies exactly rule has three important properties: it cooperates unless provoked to do otherwise (starts with 'cooperate'), it retaliates if provoked (responds to 'defect' by defecting), and it is also forgiving (immediately stops playing 'defect' if the opponent returns to cooperating). Tit for tat captures the essence of reciprocity, and its success kick-started a vast number of studies, both theoretical and empirical, on how reciprocity could perhaps explain apparently altruistic animal (and human) behaviour. A textbook discussion of the mathematics behind this idea is found in Ch. 4 of McElreath and Boyd (2007). Examples of modern developments on reciprocity in evolutionary ecology, that by now extend much beyond 'tit for tat' and the Prisoner's Dilemma, are given by Dugatkin and Reeve (1997), Sachs *et al.* (2004) and Kun *et al.* (2006).

Newer examples of simulations in evolutionary biology and ecology include examining the consequences of divorce strategies (Dubois *et al.* 2004); hierarchy formation based on fight outcomes (Dugatkin and Earley 2003); the influence of population 'buffers' on population persistence (Ferrer *et al.* 2004); finding out if details of competition between dispersers have an influence on dispersal distance (including a comparison with analytical results, and a fit to real data: McCarthy 1997); looking at adaptation in source–sink systems with gene flow (Holt *et al.* 2004); exploring the idea that parasites could manipulate hosts to disperse more, to aid the chances that the parasite infects new hosts (Lion *et al.* 2006); a comparison of analytical and simulation methods to find out how easily sympatric speciation can happen (van Doorn *et al.* 2004); and assessing ways to manage fisheries using marine reserves, given that humans quite like behaving in ways that give them economic profit (Milon 2000). As can be guessed from the diversity of topics, this list is by no means an exhaustive one. For a more thorough review and a conceptual basis of individual-based models in particular, see DeAngelis and Mooij (2005). Grimm *et al.* (2006) have recently suggested a useful protocol for standardizing the description of complex individual-based models.

Peck (2004) offers an interesting discussion as to why one should perform simulations and points out their differing role from other modelling methods. Łomnicki (1999) and Ch. 1 in McElreath and Boyd (2007) provide a critical view. Chapters 13–15 in Otto and Day (2006) and Chs. 7 and 8 in Mangel (2006) give a good introduction to dealing with stochasticity in models, including simulations but not limited to them. Many other methods presented in the preceding chapters can be adapted to stochastic scenarios, and useful starting points are Clark and Yoshimura (1993b), Bulmer (1994; Ch. 5), McNamara (1995, 1997, 1998), Benton and Grant (1999) and Sasaki and de Jong (1999). For more on causes and consequences of dispersal, see Clobert *et al.* (2001) and Bowler and Benton (2005).

Finally, simulations can be a powerful tool when examining the consequences of null hypotheses, which create powerful randomization tests for complicated biological scenarios; this is a topic beyond the scope of this book but is very useful to empiricists (Adams and Anthony 1996). For examples see Heinsohn *et al.* (1997), Mathews (2003) and Kokko *et al.* (2004).

9

Concluding remarks

where we ask which chapter you liked most (or disliked least),
and end the book with a most useful quote

We have completed our journey through seven different modelling tools in behavioural and evolutionary ecology. But the obvious question remains ... How to choose between them? Could any of the problems have been analysed with any of the methods, or is there a correct choice of a tool in each particular case?

The psychologist Abraham Maslow once quipped 'If the only tool you have is a hammer, you tend to see every problem as a nail'. There are, indeed, researchers who only use a particular method for anything they see, or perhaps (if the symptoms of the hammer syndrome are less severe) tend to seek out problems that suit a particular method. Sometimes this is because this is the only method they know well: just as when learning a foreign language or a new programming language, everything about modelling tends to get easier when done repeatedly. This can create real 'developmental inertia', which makes researchers specialize in one method over others. Reading all the chapters of this book could help to overcome this syndrome to some extent: if, for example, it is clear that fitness is frequency dependent, then a static optimality approach is not sufficient no matter how much one might have grown fond of setting derivatives of simple functions to zero.

Therefore, while one should be prudent about how many factors one considers, one should not fall into the opposite trap either: it is erroneous to think that 'additional' factors, for example frequency-dependence (Ch. 6) or an explicit formulation of density-dependence (Ch. 7), always make a model more 'special' and less generally applicable. Instead, assuming that fitness does *not* depend on density is an assumption too, even if this is less often explicitly articulated. For example, we made this tacit assumption in Chs. 2–6 but only fleetingly mentioned the density-dependent alternative in Ch. 2 (p. 37). One could in fact argue that, given the ubiquity of population regulation (Mylius and Diekmann 1995), assuming the absence of density-dependence leads to *less*-general messages. To find out how results changed if we changed the assumptions ... yes, once again: one should model it.

However, sometimes one also finds arguments that one of the methods is really superior to others, and everything else should be dismissed. I am convinced that such a view is false. Every single modelling method makes simplifications. Most of Ch. 1 is devoted to explaining why simplifications are completely essential to the modelling process, and why there is no single criterion according to which one could decide whether including or omitting one particular feature is a good or a bad thing. Because models do not investigate reality but rather the logic of an argument, they cannot as such ever lead to erroneous conclusions (unless one makes a genuine mistake in the mathematical derivation or fails to spot a bug that has crept into a computer program).

Conclusions can be totally *useless*, however. This can happen if the assumptions omit something that really does influence the biological outcome significantly, or if they focus on a too specific corner of the world, or include processes that are unlikely to happen in nature.

Consider the following exaggerated example. Letting migrant birds possess cognitive capabilities that allow them to make exact weather forecasts half a year in advance could lead to interesting adjustments of migratory routes, but such a model would have little predictive power in the real world. This example is obviously foolish, but similar blunders can happen unintentionally. For example, in a game theory model one could end up assuming silly things about how much animals have knowledge of each other's future intentions.

Therefore, an important aspect of any modelling exercise is to make sure that very odd assumptions are not made implicitly. Because all modelling methods rest on assumptions, all of them can attract criticism that can get gruelling at worst: compare Sarkar's (2005) attack on phenotypic models with Pigliucci's (2006) equally harsh views on current quantitative genetic thinking. Both authors have a point: one should never simply get used to assumptions underlying a modelling method but keep scrutinizing them. Likewise, the debate concerning how much one should simplify is unlikely to be extinguished in the near future. Population ecologists, in particular, have spent decades arguing about the relative merits of simple versus more detailed models. The 'prehistory' of this battle is superbly documented in Sharon Kingsland's book *Modeling Nature* (1995). For newer views, compare Lomnicki's (1999) arguments with those of Jones *et al.* (2005); have a look at Odenbaugh (2005) and Benton *et al.* (2006) too, and do not forget to read the classic paper by Levins (1966).

If one was to mention one big issue where different modelling tools may make different assumptions, it is the role of constraints in evolution. There is a deep sense in which models are all offspring of the same family. For example, Page and Nowak (2002) have drawn attention to the fact that very different formulations are, in the end, part of a single unified framework. While general theorems are beyond the scope of this book, glimpses of this unity can be seen here too. For example, in the beginning of Ch. 3, an optimality approach has its roots in the same soil as a quantitative genetic approach: evolution proceeds in a fitness landscape, and genes of better-performing individuals multiply; evolution stops at a local fitness peak; and the equation turns out to be exactly the same in either approach. It should be clear that if two mathematical methods do not differ in the assumptions they make, the results cannot differ either (Charlesworth 1990; Hines and Turelli 1997; Hazel *et al.* 2004). So, if there is a dramatic difference in the outcome, it must be because there is a significant difference in the assumptions regarding the biological process itself – and the matter deserves a closer look.

One way in which modelling approaches often differ in the assumptions they make is the strength of constraints on trait evolution. Typically (though not always), so-called phenotypic approaches – optimality theory together with its 'offspring' game theory – put greater faith in the power of natural selection as an optimizing agent than the genetic approaches presented in Chs. 1 and 2. Why? Because the former approach by definition looks for a phenotype that performs well under all circumstances modelled, while the latter often (but not necessarily, and not always!) assumes a simpler machinery producing adaptive responses. Take, for example, the small bird in Ch. 5, trying to survive through a cold winter day. We made the assumption that natural selection had produced a strategy that makes birds respond optimally to the set of challenges: since there is a way to respond optimally to any combination of factors such as the time of the day and body condition, natural selection is expected to mould the bird's strategy to produce the best response to these challenges. We can have some faith in this because bird populations have gone through countless winter days during the course of evolution.

But we could have, instead, assumed that there are two loci: one determining the slope of feeding effort against body condition and the other the slope of feeding effort against time of day. (Perhaps a third one should have been added, determining an intercept, in other words a baseline feeding effort.) For examples of this kind of analysis see Ch. 8 in Rice (2004). Using this option for the bird example would have created stronger constraints: with this limited set of loci, the bird is prevented from 'discovering' very complicated responses to the time of the day and body condition, even if these would have been useful. Would our conclusions have differed dramatically from those obtained in Ch. 5? This is highly unlikely. This is because the outcomes (Ch. 5) never showed any sort of weird behaviour: the predictions always showed a simple, linear-ish threshold that specified the time-dependent body condition below which the bird should start feeding. A genetic model with two slopes could have probably captured this level of complexity too. However, if we had a reason to suspect that, say, gene flow from other populations prevented optimal responses to the trade-off between starvation and predation, then an explicit genetic treatment would have been not only possible but also necessary.

There are also other cases where a dynamic optimization approach can be criticized for being too optimistic about behavioural flexibility. For example, if some states are reached extremely infrequently during evolution, it is unlikely that natural selection has the foresight to produce the best response to an essentially novel situation. Instead, if there are situations in the model that can be regarded as 'novel' or 'very rare', it makes more sense to include assumptions such as 'beyond the environmental cue value $x = 1$, responses are assumed to be equally strong as when $x = 1$' or, perhaps, 'increasingly strong but with a constant slope from $x = 1$ onwards'. Indeed, cognitive constraints can produce phenomena such as peak shifts (Mackintosh 1974; Enquist and Arak 1998), where individuals respond as if they were extrapolating from within the adaptive region of behaviour – and it is also possible that other individuals, either con- or heterospecifics, exploit such sensory biases (e.g. Ryan 1998; Bruce *et al.* 2001).

Perhaps the most essential aspect of any modelling exercise is to decide what constraints are operating. Everything depends on this. Could males behave in ways that bring them the mating advantage *without* harming the female (Ch. 2)? How accurately can barnacles sense the presence of a predator (Ch. 3)? Can zebras produce machine guns to combat lions, or not (Ch. 6, p. 139)? If they cannot, the sensible choice for a modeller is to exclude this possibility from the model too. But the correct choice for constraints is essentially an empirical question, one that cannot be answered from within the model. When modelling small birds in the winter, it is good to know that great tits really do respond to perceived predation risk and change their diurnal pattern of mass gain accordingly (Macleod *et al.* 2005).

The empirical nature of the question of constraints extends to the genetic architecture too. Even though phenotypic models typically assume greater behavioural flexibility than genetic models, this mostly seems to arise through the fact that including more detail is easier in the former – and not because genetic models could not by their nature be made equally flexible, or because tight constraints on what is possible could not be assumed in phenotypic models too. To come back once more to the example of the small bird in the winter; it is perfectly possible to create a population genetic model with two alleles (forage, or not) and with many, many loci, each taking care of producing a response to a specific state of the bird. The analysis of such a model would be quite a nightmare, however, and I strongly doubt that birds possess specific genes for behaviour in the first hour in the morning when body weight lies between 12.1 and 12.2 g.

What the birds *really* do, and how fine-tuned their responses are, depends on what kind of physiological and cognitive rules their genes manage to produce. Models cannot be used to answer this empirical question at all (although they can comment on the selective advantage of, say, developing reaction norms as opposed to genetically fixed behaviour (e.g. deWitt *et al.* 1998; Schlichting and Pigliucci 1998; Ernande and Dieckmann 2004)). But to repeat the message from Ch. 1: regardless of what we know about the underlying genetics, models can be perfectly well used to validate the logic of the argument 'it may be adaptive to refrain from eating early in the day, to be able to escape predation during the day, assuming that … ' (here the reader is expected to insert an explanation of the conditions in which this happens in the model). And, based on what we now know, it is also perfectly valid to move on and ask what happens if the foraging situation changes depending on the bird's position in the flock's dominance hierarchy; then we clearly need game theory modelling, and so on. The choice of a modelling method comes naturally: use one that offers the most convenient tools, given the currently best estimate of constraints one believes to be operating in nature.

Since we usually have to make some decisions without knowing too exactly what happens inside the creature we are interested in, it is also wise to do quite a bit more than just look at a particular set of parameter values, or keep on using a specific modelling tool only (see Hazel *et al.* (2004) for a nice example where two approaches are compared). Consequently, it is no coincidence that the most central questions in evolutionary ecology attract the attention of several modellers, each using their favourite method(s) and applying a slightly different set of assumptions. If the results coincide, we can congratulate ourselves: we have found that our predictions seem to be quite robust and do not depend on the particular simplifications inherent in each method. But if they do not, then most likely it is important for us to know more about what kinds of constraints operate in nature – is there, for example, gene flow that prevents local adaptation – to be able to predict the course of evolution. Either way, we have learnt something in the process. So, don't be shy: try different approaches and learn as you go. It is good fun!

> I had a cousin once who studied trigonometry until his whiskers drooped,
> and when he had learnt it all, a Groke came and ate him up.
> Then there he lay in the Groke's stomach, wiser at last!
> *Moominpappa's Memoirs,* Tove Jansson (1968)

Appendix: A quick guide to MATLAB

This is a quick and informal guide to MATLAB, arranged in five lessons: the basics, functions, matrix notation, truth values and 'for' loops. This gives the most basic building blocks, but MATLAB can do a lot more. To discover more, try the commands help (e.g. help plot) and lookfor (e.g. lookfor correlation). While help gives the syntax for a single command, lookfor finds all potentially relevant functions available.

Lesson 1 The very basics

The 'Mat' in MATLAB has nothing to do with maths but with matrices: the program is a 'Matrix Laboratory'. However, one does not have to worry about the details of matrix algebra: that's where the program's roots are, but it has evolved into something quite versatile. Basically MATLAB is a calculator: at any time, you can type the following in its command window

```
» 3+4
ans =
7
```

You can store answers in variables:

```
» x=3+4
x =
7
» y=x+2
y =
9
```

What makes MATLAB so powerful is that one variable, say x, can hold more than one number at a time. In other words x can be a vector or matrix of values, so lots of values can be calculated 'in one go':

```
» x=1:5
x =
1 2 3 4 5
» x.^2
ans =
1 4 9 16 25
```

This calculated the squares of all numbers from 1 to 5: the x.^2 denotes raising x to the power two. Note the full stop before the ^ symbol. This tells MATLAB to perform the calculation for each value separately. (Otherwise it tries to do something according to the multiplication rules of matrix algebra, which we won't go into here; see Otto and Day (2006).) One must get used to this: unless specifically wanting to perform matrix multiplication, one should always use full stops before * (multiplication), / (division) and ^ (power) if dealing with variables that contain more than a single value.

Sometimes we want to calculate so many values that it is not practical to look at them on the screen. You can suppress the output by adding a semicolon (;) after the line. For example, we might want to plot the square of all real numbers between 0 and 1. Why? For example: it could be that the cost of a behaviour increases more steeply as an individual spends more of its time doing it. We would like to model this, and would like to have a feel for functions that have such increasing steepness.

There is a problem. There are infinitely many real numbers between 0 and 1, and we can't possibly ask MATLAB to plot *all* of them. So let's make a plot of, say, each value from 0 to 1 with a step size of 0.01:

```
x=0:0.01:1;
```

Note the semicolon at the end. If one forgets it, there will be lots of values flowing across the screen, but apart from being tiring to watch it doesn't matter: the value of x was defined in any case.

Then we can plot the squared values:

```
y=x.^2;
plot(x,y)
```

Or we could have directly said

```
plot(x,x.^2)
```

Also, we could have specified better variable names, to help us to remember what we were thinking about:

```
activitylevel=0:0.01:1;
cost=activitylevel.^2;
% this plots cost against activitylevel
plot(activitylevel,cost)
```

Lesson 2 Functions

Firstly, a word about the command window and the workspace in MATLAB. The command window is what you 'see' in MATLAB: the series of the latest commands and their outputs. This is closely linked to the workspace, which is the collection of variables and their values currently in use. If your version of MATLAB does not show them in a separate window, type who and you will get a list of all variables that have been defined. Whos gives a longer output, giving the *sizes* of all variables (e.g. x=[2 3 4; 6 5 4] will be a matrix that has two rows and three columns, i.e. its size is 2×3).

You can use *functions* to do something with the variables in your workspace. For example, you already know the function plot from Lesson 1. Functions take *arguments* and do something with them. For example, plot draws them on the screen, but other functions could, for example, look for the maximum value of a series of numbers (useful if one wishes to maximize fitness), or find the point where something equals zero (useful if one is thinking about stabilizing selection), or do much more complicated tasks such as simulate butterfly dispersal. The nice feature of MATLAB is that it allows you to write your own functions and use them alongside MATLAB's own ones.

Functions are useful whenever you do something repeatedly. For example, plotting a graph is something that is needed so often, that programmers have already written the function (plot) for us. If they have not, the task is yours. So if you want to find out how a gene frequency equilibrium changes with an environmental parameter, you will need to solve for the equilibrium many times: once for each value of the environmental parameter. It then makes sense to write the seeking procedure as a function.

But let's start with a relatively useless example. Imagine that a student very often wants to know how much money he has got at the end of the month, if he'd like to spend quite a lot of money on beer and additionally there is a 250 € monthly rent to pay. The student has some income but it's not huge: he can only work part-time because he is participating in a really interesting modelling course. How big a salary does the student need? And what if he moved somewhere cheaper?

Functions always have the syntax

```
function [output]=functionname(arguments)
```

In our case, a function could look like this:

```
function moneyleft = beerproblem(salary,rent,        } one
   beermoney)                                           } line
% function moneyleft = beerproblem(salary,rent,        } one
   beermoney)                                           } line
% This calculates my financial situation.
% I assume a month has 30 days
moneyleft=salary-rent-30*beermoney;
```

The row with the % sign is a comment – you can write anything you like there to help you to understand the program. Because MATLAB writes the first %-rows on the screen when you type `help beerproblem`, it is useful to start by repeating the syntax of the first row. This helps to remember, by typing `help beerproblem` into MATLAB's command window, in what order the arguments should be given to the program, without having to open the file again.

The above lines should be saved in a file called `beerproblem.m`, and it should be stored where MATLAB can 'see' it (either its current working directory, or along MATLAB's 'path' of directories). Then one can try this out:

```
» help beerproblem
functionmoneyleft = beerproblem(salary,rent,beermoney)
This calculates my financial situation.
I assume a month has 30 days
```

Let's see what happens if he spends a whole lot on beer each day – here 30€:

```
» beerproblem(700,250,30)
ans =
-450
```

Oh no, there's way too little money left. What if the student took a job and earned a little more, say 1000€ a month?

```
» beerproblem(1000,250,30)
ans =
-150
```

Still not quite enough. What if he moved somewhere really cheap?

```
» beerproblem(1000,150,30)
ans =
-50
```

Still not quite enough. Maybe he'll have to reduce the amount of beer he drinks?

```
» beerproblem(1000,150,25)
ans =
100
```

Now the student is making savings each month. How much beer could he drink to get a zero balance? It must be somewhere between 25 and 30€/day, so let's try this:

```
beer=linspace(25,30,100);
savings=beerproblem(1000,150,beer);
plot(beer,savings);
grid
```

A few comments here. Linspace is another ready-made MATLAB function: it gives linearly spaced intervals (here, 100 values between 25 and 30). beerproblem is the function we have just created. Note that the arguments can be numerical values (30) or variables that contain numerical values (beer). Also, we stored the answer in a variable called savings, although the function specified that the output is called moneyleft. After using the function, there is no variable called moneyleft in MATLAB's workspace (you can type who to see which variables MATLAB currently remembers). This is perfectly acceptable, and in fact useful for two reasons. Firstly, you may want to investigate different scenarios and save the output in different variable names, for example, savings_if_bigsalaryjob, and savings_if_dontworkmuch. Secondly, the function might have to work through several steps before producing the final answer, and keep track of variables that can (and should) be forgotten once the calculation has completed. An example could be the probability of a male waterstrider mating with various females through time, when it's the final score (offspring production) that we are interested in. Having all the intermediate variables cluttering up our workspace would only confuse us.

Of course, if you are suddenly interested in intermediate results you can make them visible. This is done either by removing the semicolon at the end of the line, in which case they show on the screen during the calculation but *do not* enter the workspace, or by adding these variables to the list of output variables, in which case they do (you must give them names though):

```
function [moneyleft,MBB] = beerproblem(salary,rent,    ⎫ one
    beermoney)                                          ⎭ line
% function [moneyleft,MBB] = beerproblem(salary,rent,  ⎫ one
    beermoney)                                          ⎭ line
% This calculates my financial situation.
% I assume a month has 30 days
MBB=salary-rent;
moneyleft=MBB-30*beermoney;
```

Here MBB stands for 'money before beer consumption'. The output when asking `beerproblem(1000,150,25)` looks exactly like before: only `moneyleft` is given, MBB does not appear anywhere. Why? MATLAB as a default only gives the first output, but more can be obtained when asking politely for more:

```
» [moneyleft1000,MBB1000]=beerproblem(1000,150,25)
moneyleft1000 =
    100
MBB1000 =
    850
```

Now we have saved these particular values in variables that additionally remind ourselves what salary (1000€) we used to calculate them. In general, the output names given in the function do not constrain our choice of variable names in which the outputs are subsequently stored.

It is also perfectly valid to write program code and put it in a file without making it a function (i.e. without the function specification at the top). Such files are called *scripts*. The downside is that scripts 'see', and potentially modify, everything that is in the workspace. So if it says $x = 5$ somewhere in a script, the value of x in the workspace will be changed to 5. This can be a bummer if the script was just doing a minor task for you, and you had just been using x in the main part of your ideas to denote something totally different. My own advice is to use functions for most purposes but resort to scripts for things that need to be done only once. For example, I might have a function find_ess(temperature) that looks for the evolutionarily stable behaviour at a specific temperature, and I am writing a report, lecture or scientific publication and have decided on the final example that I'm going to use: for example, I will present a temperature range from 2 to 12 degrees. I then write a script that has the name example1.m, and contains the lines

```
temperature=linspace(2,12,100);
behaviour=find_ess(temperature)
plot(temperature,behaviour)
```

This makes it easy to check later what exact values I used to produce a specific example.

Lesson 3 What does (1,:) mean: vectors and matrices

MATLAB is a 'Matrix Laboratory', therefore everything is done by defining matrices and performing operations on them. Even the command $x = 1$ defines a matrix which happens to be a *scalar* (i.e. a single number); this is a matrix of size 1×1. You can ask what size x is by asking size(x). The first number indicates rows, the second number columns.

Square brackets are used in MATLAB to define vectors or matrices: for example,

```
y = [2 4 3]
```

You can then ask, for example, what is the second value in y, using normal brackets (not square, i.e. parentheses):

```
» y(2)
ans =
4
```

Or the last value:

```
» y(end)
ans =
3
```

Normal brackets are also used to find out subsets of a matrix: if you have defined

```
» A=[2 4 3; 5 7 2]
A =
2  4  3
5  7  2
```

then A(1,2) will give the answer 4: it is the number on row 1, column 2.

You can also ask what is A(1,2:3). MATLAB returns all the numbers on row 1, columns 2 to 3.

Then you can ask A(:,1). The colon means 'all of them'. So: let's ask for the numbers on all rows, but only when they are in column 1:

```
» A(:,1)
ans =
2
5
```

You can combine numbers in any way you want:

```
% this takes A's third column and puts it before A's first
% column
» B=[A(:,3) A(:,1)]
B =
3  2
2  5
```

And you can change individual values:

```
» B(1,1)=0
B =
0  2
2  5
```

But you cannot define something that does not form a nice rectangular matrix. If you try, you will get a rude reply from MATLAB:

```
B=[A(2,:) A(:,1)]
??? All matrices on a row in the bracketed expression
   must have the same number of rows.
```

It is up to you to decide what the rows and columns mean. For example, if you have a matrix that specifies the number of birds aged 0, 1 and 2+ years in various years, the rows could denote age while the columns indicate the year. Then Birds(:,34) could represent the population size (fledglings, 1 year olds, and older) in year 34. But note that the row numbering starts from 1, so if you decide to include age 0 birds in a matrix, it is up to you to remember that x(1,10) refers to age 0 birds in year 10, and not age 1 birds. In Box 5.1 (p. 100) we indeed had to think before indexing all rows correctly.

The numbers '1' and '10' in x(1,10) are values that index vectors or matrices. Many MATLAB functions return index values. For example, an often needed function is max, which looks for the maximum value of a vector or matrix.

Example: let's think about a trait that could vary between 0 and 2:

```
» x=0:0.01:2;
```

(The same thing can also be done using linspace, see Box 2.1.)
Fitness might be low if x is very low, but too high x is costly too:

```
» fit=x.*exp(-x);
» plot(x,fit)
```

It very much looks like $x = 1$ is the optimum. But is it exactly there? Let's check:

```
» [best,bestindex]=max(fit)
best =
0.3679
bestindex =
101
```

This tells us that the best possible fitness is 0.3679, and it is found at the 101st value of all the fitness values we calculated. So what is the best value of x? Because we calculated one fitness value for each value of x, and the 101st value of fitness is highest, the optimal x is also the 101st value of x.

```
» x(bestindex)
ans =
1
```

So yes, x=1 is optimal.

Lesson 4 Truth and lies in MATLAB

Often, we need values that tell us if something is true or not. For example, in Ch. 5 we asked if it is better to forage than to rest.

'True' in MATLAB is the value 1. 'False' is the value 0. So, for example, you can ask:

```
» 2>3
ans =
0
```

This may look weird, but it simply means that $2 > 3$ is not true. You can also ask the same thing for multiple values. For example, you might have calculated the fitness associated with behaviour A or B, in situations 1, 2 or 3 (whatever they are, e.g. 'predator present', 'not present', 'not sure'). You have obtained the result

```
» fitnessA=[2.3 3.4 3.3];
» fitnessB=[0.4 3.6 6.7];
```

Then you want to know when A is better than B:

```
» fitnessA>fitnessB
ans =
1 0 0
```

So in the first case it is good to do A, while in the other two cases the individual should not do A (and do B instead).

Next we will meet the operator ~, which means 'not'. One might want to form a vector BestFitness to indicate how high the fitness of an individual can get if it always chooses the optimal tactic. BestFitness should get its value from fitnessA when fitnessA>fitnessB, otherwise it should get its value from fitnessB. How to do this? The trick is to realize that this is a weighted sum of fitnessA and fitnessB. The weight should be 1 for A and 0 for B when $A > B$, and 0 for A and 1 for B otherwise.

How to get the weights? We already have them above: they are the answer to the question fitnessA>fitnessB. So we can type

```
» DoA=fitnessA>fitnessB
DoA =
1 0 0
» BestFitness=DoA.*fitnessA + ~DoA.*fitnessB
```

```
BestFitness =
2.3000  3.6000  6.7000
```

Note the ~DoA, which means 'don't do A', or more accurately, 'weight equals 1 when A is not true, 0 when A is true'.

Lesson 5 Loops

What does this mean:

```
for i=1:10
  double(i)=i*2;
  half(i)=i/2;
end;
```

The loop does something for all the values of *i*. In other words, it goes through all the values of *i* from 1 to 10, and updates two variables double and half along the way. The *i*th value of double becomes 2*i, and the *i*th value of half becomes i/2. The results: double has ten values from 2 to 20, and half has values from 0.5 to 5.

Of course, this is a silly example. It is far quicker to define double=2:2:20 and half=0.5:0.5:5 without doing it using a complicated loop. Also, you would avoid the problem that if some values, say 100 different ones, already exist in the variable double the loop only redefines the 10 first ones and you might end up being quite confused why there are more. If you encounter these sort of problems you ought to initialize the variables:

```
double=[ ] % makes an empty variable
```

or

```
double=zeros([1 10]) % makes a matrix of 1 row, 10     } one
   columns, where every value is a zero                } line
```

before you start. (The latter is faster, by the way, because the size of the vector double does not have to be expanded every time inside the loop.)

Loops in general are quite slow to run. They perform repeated calculations one by one, when the task could be vectorized, which means dealing with the whole vector in one go, as in double=2:2:20. But sometimes loops really are needed: one cannot calculate everything simultaneously. A typical example in ecology is when something happens over time, and one cannot know what, for example, the size of the population is at time *t*+1 before its size at time *t* has been calculated. Then one will need to consider all the time steps separately:

```
N=100; % initially we have 100 individuals
for t=1:100 % calculate new population sizes
    % growth rate is a normally distributed random number,
    % mean zero, variance 1; in Matlab this is given by
    % 'randn'
    r(t)=randn(1);
    lambda(t)=exp(r(t));
    N(t+1)=N(t)*lambda(t);
end
figure(1); plot(1:101,N);
figure(2); hist(lambda,20);
```

Note the following.

- When t runs from 1 to 100, and we calculate N(t+1) for each t, N will have 101 values when we are done, not 100.
- We cannot say plot(t,N) at the end. That is because t does not contain all the values from 1 to 100 (or 101). It contains only the last value, t=100.
- hist is a MATLAB function that plots histograms of a variable, in this case putting the data (lambda) in 20 different categories.
- It would be good if we had set r and lambda to be empty in the beginning. The lines work without this setting, but there will be problems if they contain previous data and one does not recalculate things for all 100 values to replace the previous ones: for example if one changes t=1:100 to t=1:20. Functions (Lesson 2) would be useful here.

But what if one wants to run a loop for values other than integers? For example, you might want to solve the optimal behaviour for 100 different conditions of an animal, instead of 100 different integer-valued years. You cannot use the above procedure directly because

```
for condition=0:0.01:1
```

cannot be followed by

```
behaviour(condition)= something %some calculation   } one
    here                                              } line
```

Why not? Because behaviour(condition)=something will start in the loop with behaviour(0)=something, and one cannot define the zeroeth value of a matrix, and the 0.01th value is equally ill defined. Instead, do the following:

```
condition=0:0.01:1 % these are the ones to be calculated
  for i=1:length(condition)
  behaviour(i)= % insert some calculation that uses    } one
    condition(i)                                        } line
end;
```

Now you can plot

```
plot(condition,behaviour)
```

without any problems: both `condition` and `behaviour` will be vectors that are identical in size. (Of course you will have to define the calculation first.)

Loops can be nested inside each other. For example:

```
condition=0:0.01:1
genotype=[1 2]
for i=1:length(condition)
  for j=1:length(genotype)
    fitness(i,j)=genotype(j)*condition(i);
  end;
end;
```

What does this do? It creates a 101×2 sized matrix of fitness values where fitness depends both on condition and genotype, with genotype 2 having double the fitness of genotype 1. This is not the quickest way to do things, since vectorizing will always work better, but it is perhaps the simplest to start with.

The vectorized way would be both shorter to write and faster to run:

```
condition=(0:0.01:1)';
genotype=[1  2];
fitness=condition*genotype;
```

Note that we now did not write a dot before the multiplication sign. This is because the vectorized form takes advantage of matrix algebra, which offers ways to multiply vectors to yield a matrix of a specific size. A full treatment of vectorization methods is beyond the scope of this book as we have not covered matrix algebra; for a starting point see Ch. 7 in Rice (2004).

References

Adams, D. C. and Anthony, C. D. (1996). Using randomization techniques to analyse behavioural data. *Animal Behaviour* 51, 733–738.

Andersson, M. (1982). Female choice selects for extreme tail length in a widowbird. *Nature* 299, 818–820.

Andersson, M. (1994). *Sexual Selection*. Princeton: Princeton University Press.

Andersson, M. and Simmons, L. W. (2006). Sexual selection and mate choice. *Trends in Ecology and Evolution* 21, 296–302.

Arkes, H. R. and Ayton, P. (1999). The sunk cost and Concorde effects: are humans less rational than lower animals? *Psychological Bulletin* 125, 591–600.

Arnold, S. J. (1992). Constraints on phenotypic evolution. *American Naturalist* 140, S85–S108.

Arnqvist, G. (2006). Sensory exploitation and sexual conflict. *Philosophical Transactions of the Royal Society of London, Series B* 361, 375–386.

Arnqvist, G. and Rowe, L. (2005). *Sexual Conflict*. Princeton: Princeton University Press.

Aspi, J., Jäkäläniemi, A., Tuomi, J. and Siikamäki, P. (2003). Multilevel phenotypic selection on morphological characters in a metapopulation of *Silene tatarica*. *Evolution* 57, 509–517.

Axelrod, R. and Hamilton, W. D. (1981). The evolution of cooperation. *Science* 211, 1390–1396.

Bakker, T. C. M. (1999). The study of intersexual selection using quantitative genetics. *Behaviour* 136, 1237–1265.

Baskett, M. L., Levin, S. A., Gaines, S. D. and Dushoff, J. (2005). Marine reserve design and the evolution of size at maturation in harvested fish. *Ecological Applications* 15, 882–901.

Bednekoff, P. A. and Houston, A. I. (1994). Avian daily foraging patterns: effects of digestive constraints and variability. *Evolutionary Ecology* 8, 36–52.

Benton, T. G. and Grant, A. (1999). Optimal reproductive effort in stochastic, density-dependent environments. *Evolution* 53, 677–688.

Benton, T. G., Plaistow, S. J. and Coulson, T. N. (2006). Complex population dynamics and complex causation: devils, details and demography. *Proceedings of the Royal Society of London, Series B* 273, 1173–1181.

Berg, F., Gustafson, U. and Andersson, L. (2006). The uncoupling protein 1 gene (UCP1) is disrupted in the pig lineage: a genetic explanation for poor thermoregulation in piglets. *PLoS Genetics* 2, 1178–1181.

Björklund, M. (2004). Constancy of the G matrix in ecological time. *Evolution* 58, 1157–1164.

Blows, M. W. and Hoffmann, A. A. (2005). A reassessment of genetic limits to evolutionary change. *Ecology* 86, 1371–1384.

Boake, C. R. B. (1994). *Quantitative Genetic Studies of Behavioral Evolution*. Chicago: Chicago University Press.

Borges, J. L. (1975). *A Universal History of Infamy*. London: Penguin Books.

Bowler, D. E. and Benton, T. G. (2005). Causes and consequences of animal dispersal strategies: relating individual behaviour to spatial dynamics. *Biological Reviews* 80, 205–225.

Brandt, L. S. E. and Greenfield, M. D. (2004). Condition-dependent traits and the capture of genetic variance in male advertisement song. *Journal of Evolutionary Biology* 17, 821–828.

Brodin, A. (2000). Why do hoarding birds gain fat in winter in the wrong way? Suggestions from a dynamic model. *Behavioral Ecology* 11, 27–39.

Brodin, A. (2007). Theoretical models of adaptive energy management in small wintering birds. *Philosophical Transactions of the Royal Society London, Series B*, in press.

Brommer, J. E. (2000). The evolution of fitness in life-history theory. *Biological Reviews* 75, 377–404.

Brommer, J. E., Merilä, J. and Kokko, H. (2002). Reproductive timing and individual fitness. *Ecology Letters* 5, 802–810.

Brommer, J. E., Gustafsson, L., Pietiäinen, H. and Merilä, J. (2004). Single-generation estimates of individual fitness as proxies for long-term genetic contribution. *American Naturalist* 163, 505–517.

Bruce, M. J., Herberstein, M. E. and Elgar, M. A. (2001). Signalling conflict between prey and predator attraction. *Journal of Evolutionary Biology* 14, 786–794.

Bulmer, M. (1994). *Theoretical Evolutionary Ecology*. Sunderland: Sinauer.

Butlin, R. K. and Tregenza, T. (2005). The way the world might be. *Journal of Evolutionary Biology* 18, 1205–1208.

Camerer, C. F. (2003). *Behavioral Game Theory: Experiments in Strategic Interaction*. Princeton: Princeton University Press.

Candolin, U. (1999). The relationship between signal quality and physical condition: is sexual signalling honest in the three-spined stickleback? *Animal Behaviour* 58, 1261–1267.

Candolin, U. (2000). Increased signalling effort when survival prospects decrease: male–male competition ensures honesty. *Animal Behaviour* 60, 417–422.

Case, T. J. (2000). *An Illustrated Guide to Theoretical Ecology*. Oxford: Oxford University Press.

Chapman, T., Arnqvist, G., Bangham, J. and Rowe, L. (2003). Sexual conflict. *Trends in Ecology and Evolution* 18, 41–47.

Charlesworth, B. (1990). Optimization models, quantitative genetics and mutation. *Evolution* 44, 520–538.

Christiansen, F. B. (1999). *Population Genetics of Multiple Loci*. Chichester, UK: John Wiley.

Claessen, D., de Roos, A. M. and Persson, L. (2004). Population dynamic theory of size-dependent cannibalism. *Proceedings of the Royal Society of London, Series B* 271, 333–340.

Clark, C. W. and Ekman, J. (1995). Dominant and subordinate fattening strategies: a dynamic game. *Oikos* 72, 205–212.

Clark, C. W. and Mangel, M. (2000). *Dynamic State Variable Models in Ecology: Methods and Applications*. Oxford: Oxford University Press.

Clark, C. W. and Yoshimura, J. (1993a). Behavioral responses to variations in population size: a stochastic evolutionary game. *Behavioral Ecology* 4, 282–288.

Clark, C. W. and Yoshimura, J. (1993b). Optimization and ESS analysis for populations in stochastic environments. In *Lecture Notes in Biomathematics 98: Adaptation in Stochastic Environments*, eds. J. Yoshimura and C. W. Clark, pp. 122–131. Berlin: Springer-Verlag.

Clobert, J., Danchin, E., Dhondt, A. A. and Nichols, J. D. (eds.) (2001). *Dispersal*. Oxford: Oxford University Press.

Comins, H. N., Hamilton, W. D. and May, R. M. (1980). Evolutionarily stable dispersal strategies. *Journal of Theoretical Biology* 82, 205–230.

Coppack, T. and Both, C. (2002). Predicting life-cycle adaptation of migratory birds to global climate change. *Ardea* 90, 369–378.

Cotton, S., Fowler, K. and Pomiankowski, A. (2004). Condition-dependence of sexual ornament size and variation in the stalk-eyed fly *Cyrtodiopsis dalmanni* (Diptera: Diopsidae). *Evolution* 58, 1038–1046.

Courchamp, F., Clutton-Brock, T. and Grenfell, B. (1999). Inverse density dependence and the Allee effect. *Trends in Ecology and Evolution* 10, 405–410.

Creel, S. R. (1990a). How to measure indirect fitness. *Proceedings of the Royal Society of London, Series B* 241, 229–231.

Creel, S. R. (1990b). The future components of inclusive fitness: accounting for interactions between members of overlapping generations. *Animal Behaviour* 40, 127–134.

Crudgington, H. S. and Siva-Jothy, M. T. (2000). Genital damage, kicking and early death: the battle of the sexes takes a sinister turn in the bean weevil. *Nature* 407, 855–856.

Curio, E. (1987). Animal decision-making and the concorde fallacy. *Trends in Ecology and Evolution* 2, 148–152.

Dawkins, R. (1976). *The Selfish Gene*. Oxford: Oxford University Press.

Day, T. and Taylor, P. D. (1997). Hamilton's rule meets the Hamiltonian: kin selection on dynamic characters. *Proceedings of the Royal Society of London, Series B* 264, 639–644.

DeAngelis, D. L. and Mooij, W. M. (2005). Individual-based modeling of ecological and evolutionary processes. *Annual Reviews of Ecology, Evolution and Systematics* 36, 147–168.

DeWitt, T. J., Sih, A. and Wilson, D. S. (1998). Costs and limits of phenotypic plasticity. *Trends in Ecology and Evolution* 13, 77–81.

Dieckmann, U. (1997). Can adaptive dynamics invade? *Trends in Ecology and Evolution* 12, 128–131.

Dieckmann, U. and Metz, J. A. J. (2006). Surprising evolutionary predictions from enhanced ecological realism. *Theoretical Population Biology* 69, 263–281.

Doherty, P. F. Jr, Sorci, G., Royle, J. A. *et al.* (2003). Sexual selection affects local extinction and turnover in bird communities. *Proceedings of the National Academy of Sciences USA* 100, 5858–5862.

Dubois, F., Wajnberg, É. and Cézilly, F. (2004). Optimal divorce and re-mating strategies for monogamous female birds: a simulation model. *Behavioral Ecology and Sociobiology* 56, 228–236.

Dugatkin, L. A. (1997). *Cooperation Among Animals.* Oxford: Oxford University Press.

Dugatkin, L. A. and Earley, R. L. (2003). Group fusion: the impact of winner, loser, and bystander effects on hierarchy formation in large groups. *Behavioral Ecology* 14, 367–373.

Dugatkin, L. A. and Reeve, H. K. (eds.) (1998). *Game Theory and Animal Behavior.* Oxford: Oxford University Press.

Eadie, J. M. and Fryxell, J. M. (1992). Density dependence, frequency dependence, and alternative nesting strategies in goldeneyes. *American Naturalist* 140, 621–641.

Edvardsson, M. and Tregenza, T. (2005). Why do male *Callosobruchus maculatus* harm their mates? *Behavioral Ecology* 16, 788–793.

Ehrlich, P. R. and Hanski, I. (eds.) (2004). *On the Wings of Checkerspots: A Model System for Population Biology.* Oxford: Oxford University Press.

Enquist, M. and Arak, A. (1998). Neural representation and the evolution of signal form. In *Cognitive Ecology*, ed. R. Dukas, pp. 21–87. Chicago: University of Chicago Press.

Ens, B. J., Weissing, F. J. and Drent, R. H. (1995). The despotic distribution and deferred maturity: two sides of the same coin. *American Naturalist* 146, 625–650.

Ernande, B. and Dieckmann, U. (2004). The evolution of phenotypic plasticity in spatially structured environments: implications of intraspecific competition, plasticity costs and environmental characteristics. *Journal of Evolutionary Biology* 17, 613–628.

Eshel, I. (1982). Evolutionarily stable strategies and viability selection in Mendelian populations. *Theoretical Population Biology* 22, 204–217.

Eshel, I. and Feldman, M. W. (2001). Optimality and evolutionary stability under short-term and long-term selection. In *Adaptationism and Optimality*, eds. S. H. Orzack and E. Sober, pp. 161–190. Cambridge: Cambridge University Press.

Falconer, D. S. and Mackay, T. F. C. (1996). *Introduction to Quantitative Genetics.* 4th edn. Harlow: Longman.

Falster, D. S. and Westoby, M. (2003). Plant height and evolutionary games. *Trends in Ecology and Evolution* 18, 337–343.

Ferrer, M., Otalora, F. and Garca-Ruiz, J. M. (2004). Density-dependent age of first reproduction as a buffer affecting persistence of small populations. *Ecological Applications* 14, 616–624.

Ferrière, R., Dieckmann, U. and Couvet, D. (eds.) (2004a). *Evolutionary Conservation Biology*. Cambridge: Cambridge University Press.

Ferrière, R., Dieckmann, U. and Couvet, D. (2004b). Introduction. In *Evolutionary Conservation Biology*, eds. R. Ferrière, U. Dieckmann, and D. Couvet, pp. 1–14. Cambridge: Cambridge University Press.

Fisher, R. A. (1930). *The Genetical Theory of Natural Selection*. Oxford: Oxford University Press.

Frank, R. (1999). *Luxury Fever*. Princeton: Princeton University Press.

Frank, S. A. (1998). *Foundations of Social Evolution*. Princeton: Princeton University Press.

Gandon, S. and Michalakis, Y. (1999). Evolutionarily stable dispersal rate in a metapopulation with extinctions and kin competition. *Journal of Theoretical Biology* 199, 275–290.

Gardner, A. and West, S. A. (2004). Cooperation and punishment, especially in humans. *American Naturalist* 164, 753–764.

Gavrilets, S. and Rice, W. R. (2006). Genetic models of homosexuality: generating testable predictions. *Proceedings of the Royal Society of London, Series B* 273, 3031–3038.

Gavrilets, S., Arnquist, G. and Friberg, U. (2001). The evolution of female mate choice by sexual conflict. *Proceedings of the Royal Society of London, Series B* 268, 531–539.

Getty, T. (2006). Sexually selected signals are not similar to sports handicaps. *Trends in Ecology and Evolution* 21, 83–88.

Gill, J. A., Norris, K., Potts, P. M. *et al.* (2001). The buffer effect and large-scale population regulation in migratory birds. *Nature* 412, 436–438.

Gillespie, J. H. (1998). *Population Genetics: A Concise Guide*. Baltimore: Johns Hopkins University Press.

Givnish, T. J. (1982). Adaptive significance of leaf height in forest herbs. *American Naturalist* 120, 353–381.

Godfray, H. C. J. and Parker, G. A. (1991). Clutch size, fecundity and parent–offspring conflict. *Philosophical Transactions of the Royal Society of London, Series B* 332, 67–80.

Gomulkiewicz, R. (1998). Game theory, optimization, and quantitative genetics. In *Game Theory and Animal Behavior*, eds. L. A. Dugatkin and H. K. Reeve, pp. 283–303. Oxford: Oxford University Press.

Gould, S. J. and Lewontin, R. C. (1979). The Spandrels of San Marco and the Panglossian paradigm: a critique of the adaptationist programme. *Proceedings of the Royal Society of London, Series B* 205, 581–598.

Grimm, V., Berger, U., Bastiansen, F., *et al.* (2006). A standard protocol for describing individual-based and agent-based models. *Ecological Modelling* 198, 115–126.

Gyllenberg, M., Parvinen, K. and Dieckmann, U. (2002). Evolutionary suicide and evolution of dispersal in structured metapopulations. *Journal of Mathematical Biology* 45, 79–105.

Hamilton, W. D. (1964). The genetical evolution of social behaviour, I and II. *Journal of Theoretical Biology* 7, 1–52.

Hamilton, W. D. and May, R. M. (1977). Dispersal in stable habitats. *Nature* 269, 578–581.

Hanski, I. (1998). Metapopulation dynamics. *Nature* 396, 41–49.

Hanski, I. (1999). *Metapopulation Ecology*. Oxford: Oxford University Press.

Härdling, R. and Kaitala, A. (2005). The evolution of repeated mating under sexual conflict. *Journal of Evolutionary Biology* 18, 106–115.

Hazel, W. N., Smock, R. and Johnson, M. D. (1990). A polygenic model for the maintenance and evolution of conditional strategies. *Proceedings of the Royal Society of London, Series B* 242, 181–187.

Hazel, W., Smock, R. and Lively, C. M. (2004). The ecological genetics of conditional strategies. *American Naturalist* 163, 888–900.

Heino, M. and Hanski, I. (2001). Evolution of migration rate in a spatially realistic metapopulation model. *American Naturalist* 157, 495–511.

Heinsohn, R., Legge, S. and Barry, S. (1997). Extreme bias in sex allocation in *Eclectus* parrots. *Proceedings of the Royal Society of London, Series B* 264, 1325–1329.

Hines, W. G. S. and Turelli, M. (1997). Multilocus evolutionary stable strategy models: additive effects. *Journal of Theoretical Biology* 187, 379–388.

Hoelzer, G. A. (1989). The good parent process of sexual selection. *Animal Behaviour* 38, 1067–1078.

Holland, B. and Rice, W. R. (1998). Chase-away selection: antagonistic seduction vs. resistance. *Evolution* 52, 1–7.

Holt, R. D. and Gomulkiewicz, R. (2004). Conservation implications of niche conservatism and evolution in heterogeneous environments. In *Evolutionary Conservation Biology*, eds. R. Ferrière, U. Dieckmann and D. Couvet, pp. 244–264. Cambridge: Cambridge University Press.

Holt, R. D., Knight, T. M. and Barfield, M. (2004). Allee effects, immigration, and the evolution of species' niches. *American Naturalist* 163, 253–262.

Houston, A. I. and McNamara, J. M. (1999). *Models of Adaptive Behaviour: An Approach Based on State*. Cambridge: Cambridge University Press.

Houston, A. I. and McNamara, J. M. (2005). John Maynard Smith and the importance of consistency in evolutionary game theory. *Biology and Philosophy* 20, 933–950.

Houston, A. I., Székely, T. and McNamara, J. M. (2005). Conflict between parents over care. *Trends in Ecology and Evolution* 20, 33–38.

Hunt, J., Brooks, R., Jennions, M. D. *et al.* (2004). High-quality male field crickets invest heavily in sexual display but die young. *Nature* 432, 1024–1027.

Iwasa, Y., Pomiankowski, A. and Nee, S. (1991). The evolution of costly mate preferences. II. The 'handicap' principle. *Evolution* 45, 1431–1442.

Johnstone, R. A. and Keller, L. (2000). How males gain by harming their mates: sexual conflict, seminal toxins, and the cost of mating. *American Naturalist* 156, 368–377.

Jones, A. G., Arnold, S. J. and Borger, R. (2003). Stability of the G-matrix in a population experiencing pleiotropic mutation, stabilizing selection, and genetic drift. *Evolution* 57, 1747–1760.

Jones, O. R., Crawley, M. J., Pilkington, J. G. and Pemberton, J. M. (2005). Predictors of early survival in Soay sheep: cohort-, maternal- and individual-level variation. *Proceedings of the Royal Society of London, Series B* 272, 2619–2625.

Jönsson, K. I., Tuomi, J. and Järemo, J. (1998). Pre- and postbreeding costs of parental investment. *Oikos* 83, 424–431.

Kahneman, D., Krueger, A. B., Schkade, D., Schwarz, N. and Stone, A. A. (2006). Would you be happier if you were richer? A focusing illusion. *Science* 312, 1908–1910.

Kaitala, A., Kaitala, V. and Lundberg, P. (1993). A theory of partial migration. *American Naturalist* 142, 59–81.

Kemp, D. J. and Wiklund, C. (2004). Residency effects in animal contests. *Proceedings of the Royal Society of London, Series B* 271, 1707–1711.

Kingsland, S. E. (1995). *Modeling Nature*, 2nd edn Chicago: University of Chicago Press.

Kirkpatrick, M. and Barton, N. H. (1997). Evolution of a species' range. *American Naturalist* 150, 1–23.

Koch, G. W., Sillett, S. C., Jennings, G. M. and Davis, S. D. (2004). The limits to tree height. *Nature* 428, 851–854.

Koenig, W. D. and Walters, J. R. (1999). Sex-ratio selection with helpers at the nest: the repayment model revisited. *American Naturalist* 153, 124–130.

Kokko, H. (1997). Evolutionarily stable strategies of age-dependent sexual advertisement. *Behavioral Ecology and Sociobiology* 41, 99–107.

Kokko, H. and Brooks, R. (2003). Sexy to die for? Sexual selection and the risk of extinction. *Annales Zoologici Fennici* 40, 207–219.

Kokko, H. and Jennions, M. (2003). It takes two to tango. *Trends in Ecology and Evolution* 18, 103–104.

Kokko, H. and López-Sepulcre, A. (2006). From individual dispersal to species ranges: perspectives for a changing world. *Science* 313, 789–791.

Kokko, H. and Lundberg, P. (2001). Dispersal, migration and offspring retention in saturated habitats. *American Naturalist* 157, 188–202.

Kokko, H. and Rankin, D. J. (2006). Lonely hearts or sex in the city? Density-dependent effects in mating systems. *Philosophical Transactions of the Royal Society of London, Series B* 361, 319–334.

Kokko, H. and Sutherland, W. J. (1998). Optimal floating and queuing strategies: consequences for density dependence and habitat loss. *American Naturalist* 152, 354–366.

Kokko, H., Brooks, R., Jennions, M. and Morley, J. (2003). The evolution of mate choice and mating biases. *Proceedings of the Royal Society of London, Series B* 270, 653–664.

Kokko, H., Harris, M. P. and Wanless, S. (2004). Competition for breeding sites and site-dependent population regulation in a highly colonial seabird, the common guillemot *Uria aalge*. *Journal of Animal Ecology* 73, 367–376.

Kokko, H., Jennions, M. D. and Brooks, R. (2006a). Unifying and testing models of sexual selection. *Annual Reviews of Ecology, Evolution and Systematics* 37, 43–66.

Kokko, H., López-Sepulcre, A. and Morrell, L. J. (2006b). From hawks and doves to self-consistent games of territorial behavior. *American Naturalist* 167, 901–912.

Kölliker, M., Brodie, E. D. and Moore, A. J. (2005). The coadaptation of parental supply and offspring demand. *American Naturalist* 166, 506–516.

Kotiaho, J. S. (2007). The stability of genetic variance–covariance matrix in the presence of selection. *Journal of Evolutionary Biology* 20, 28–29.

Kotiaho, J., Alatalo, R. V., Mappes, J., Parri, S. and Rivero, A. (1998). Male mating success and risk of predation in a wolf spider: a balance between sexual and natural selection? *Journal of Animal Ecology* 67, 287–291.

Kotiaho, J. S., Simmons, L. W. and Tomkins, J. L. (2001). Towards a resolution of the lek paradox. *Nature* 410, 684–686.

Krams, I. (2002). Mass-dependent take-off ability in wintering great tits (*Parus major*): comparison of top-ranked adult males and subordinate juvenile females. *Behavioral Ecology and Sociobiology* 51, 345–349.

Kriska, G., Horvath, G. and Andrikovics, S. (1998). Why do mayflies lay their eggs en masse on dry asphalt roads? Water-imitating polarized light reflected from asphalt attracts Ephemeroptera. *Journal of Experimental Biology* 201, 2273–2286.

Kullberg, C., Fransson, T. and Jakobsson, S. (1996). Impaired predator evasion in fat blackcaps (*Sylvia atricapilla*). *Proceedings of the Royal Society of London, Series B* 263, 1671–1675.

Kun, Á., Boza, G. and Scheuring, I. (2006). Asynchronous snowdrift game with synergistic effect as a model of cooperation. *Behavioral Ecology* 17, 633–641.

Laland, K. N., Odling-Smee, F. J. and Feldman, M. W. (1996). The evolutionary consequences of niche construction: a theoretical investigation using two-locus theory. *Journal of Evolutionary Biology* 9, 293–316.

Lande, R. (1982). A quantitative genetic theory of life history evolution. *Ecology* 63, 607–615.

Lehmann, L. and Keller, L. (2006). The evolution of cooperation and altruism: a general framework and a classification of models. *Journal of Evolutionary Biology* 19, 1365–1376.

Lessells, C. M. (2005). Why are males bad for females? Models for the evolution of damaging male mating behavior. *American Naturalist* 165, S46–S63.

Lessells, C. M. (2006). The evolutionary outcome of sexual conflict. *Philosophical Transactions of the Royal Society of London, Series B* 361, 301–317.

Levins, R. (1966). The strategy of model building in population biology. *American Scientist* 54, 421–431.

Lima, S. and Dill, L. M. (1990). Behavioral decisions made under the risk of predation: a review and prospects. *Canadian Journal of Zoology* 68, 619–640.

Lind, J., Fransson, T., Jakobsson, S. and Kullberg, C. (1999). Reduced take-off ability in robins (*Erithacus rubecula*) due to migratory fuel load. *Behavioral Ecology and Sociobiology* 46, 65–70.

Lion, S., van Baalen, M. and Wilson, W. G. (2006). The evolution of parasite manipulation of host dispersal. *Proceedings of the Royal Society of London, Series B* 273, 1063–1071.

Lively, C. M. (1986a). Competition, comparative life histories, and maintenance of shell dimorphism in a barnacle. *Ecology* 67, 858–864.

Lively, C. M. (1986b). Predator-induced shell dimorphism in the acorn barnacle *Chthamalus anisopoma*. *Evolution* 40, 232–242.

Łomnicki, A. (1999). Individual-based models and the individual-based approach to population ecology. *Ecological Modelling* 115, 191–198.

Lorch, P. D., Proulx, S., Rowe, L. and Day, T. (2003). Condition-dependent sexual selection can accelerate adaptation. *Evolutionary Ecology Research* 5, 867–881.

Lynch, M. and Walsh, B. (1998). *Genetics and Analysis of Quantitative Traits*. Sunderland: Sinauer.

Mackintosh, N. J. (1974). *The Psychology of Animal Learning*. London: Academic Press.

Macleod, R., Gosler, A. G. and Cresswell, W. (2005). Diurnal mass gain strategies and perceived predation risk in the great tit *Parus major*. *Journal of Animal Ecology* 74, 956–964.

Mangel, M. (2006). *The Theoretical Biologist's Toolbox*. Cambridge: Cambridge University Press.

Mangel, M., Rosenheim, J. A. and Adler, F. R. (1994). Clutch size, offspring performance, and intergenerational fitness. *Behavioral Ecology* 5, 412–417.

Marrow, P., Johnstone, R. A. and Hurst, L. D. (1996). Riding the evolutionary streetcar: where population genetics and game theory meet. *Trends in Ecology and Evolution* 11, 445–446.

Matthews, L. M. (2003). Tests of the male-guarding hypothesis for social monogamy: male snapping shrimps prefer to associate with high-value females. *Behavioral Ecology* 14, 63–67.

Maynard Smith, J. (1977). Parental investment: a prospective analysis. *Animal Behaviour* 25, 1–9.

Maynard Smith, J. (1982). *Evolution and the Theory of Games*. Cambridge: Cambridge University Press.

McCarthy, M. A. (1997). Competition and dispersal from multiple nests. *Ecology* 78, 873–883.

McElreath, R. and Boyd, R. (2007). *Mathematical Models of Social Evolution: A Guide for the Perplexed*. Chicago: Chicago University Press.

McNamara, J. M. (1997). Optimal life histories for structured populations in fluctuating environments. *Theoretical Population Biology* 51, 94–108.

McNamara, J. M. (1998). Phenotypic plasticity in fluctuating environments: consequences of the lack of individual optimization. *Behavioral Ecology* 9, 642–648.

McNamara, J. M. and Houston, A. I. (1986). The common currency for behavioural decisions. *American Naturalist* 127, 358–378.

McNamara, J. M., Webb, J. N. and Collins, E. J. (1995). Dynamic optimization in fluctuating environments. *Proceedings of the Royal Society of London, Series B* 261, 279–284.

McNamara, J. M., Welham, R. K. and Houston, A. I. (1998). The timing of migration within the context of an annual routine. *Journal of Avian Biology* 29, 416–423.

McNamara, J. M., Houston, A. I. and Collins, E. J. (2001). Optimality models in behavioral biology. *SIAM Review* 43, 413–466.

Mead, L. S. and Arnold, S. J. (2004). Quantitative genetic models of sexual selection. *Trends in Ecology and Evolution* 19, 264–271.

Mesterton-Gibbons, M. (2000). *An Introduction to Game-theoretic Modelling*, 2nd edn. Rhode Island: American Mathematical Society.

Meszéna, G., Kisdi, É., Dieckmann, U., Geritz, S. A. H. and Metz, J. A. J. (2001). Evolutionary optimization methods and matrix games in the unified perspective of adaptive dynamics. *Selection* 2, 193–210.

Metcalfe, N. B. and Ure, S. E. (1995). Diurnal variation in flight performance and hence potential predation risk in small birds. *Proceedings of the Royal Society of London, Series B* 261, 395–400.

Metz, J. A. J., Nisbet, R. M. and Geritz, S. A. H. (1992). How should we define 'fitness' for general ecological scenarios? *Trends in Ecology and Evolution* 7, 198–202.

Milon, J. W. (2000). Pastures, fences, tragedies and marine reserves. *Bulletin of Marine Science* 66, 901–916.

Moore, A. J. and Boake, C. R. B. (1994). Optimality and evolutionary genetics: complementary procedures for evolutionary analysis in behavioural ecology. *Trends in Ecology and Evolution* 9, 69–72.

Morrow, E. H. and Fricke, C. (2004). Sexual selection and the risk of extinction in mammals. *Proceedings of the Royal Society of London, Series B* 271, 2395–2401.

Morrow, E. H. and Pitcher, T. E. (2003). Sexual selection and the risk of extinction in birds. *Proceedings of the Royal Society of London, Series B* 270, 1793–1799.

Moya-Laraño, J., Orta-Ocaña, J. M., Barrientos, J. A., Bach C. and Wise, D. H. (2002). Territoriality in a cannibalistic burrowing wolf spider. *Ecology* 83, 356–361.

Mueller, L. D. and Rose, M. R. (1996). Evolutionary theory predicts late-life mortality plateaus. *Proceedings of the National Academy of Sciences USA* 93, 15249–15253.

Mylius, S. D. and Diekmann, O. (1995). On evolutionarily stable life histories, optimization and the need to be specific about density dependence. *Oikos* 74, 218–224.

Nakajima, M., Matsuda, H. and Hori, M. (2005). A population genetic model for lateral dimorphism frequency in fishes. *Population Ecology* 47, 83–90.

Neumann, D. R. (1999). Agonistic behavior in harbor seals (*Phoca vitulina*) in relation to the availability of haulout space. *Marine Mammal Science* 15, 507–525.

Nowak, M. A. and Sigmund, K. (2004). Evolutionary dynamics of biological games. *Science* 303, 793–799.

Nunney, L. and Campbell, K. A. (1993). Assessing minimum viable population size: demography meets population genetics. *Trends in Ecology and Evolution* 8, 234–239.

Nussey, D. H., Postma, E., Gienapp, P. and Visser, M. E. (2005). Selection on heritable phenotypic plasticity in a wild bird population. *Science* 310, 304–306.

Odenbaugh, J. (2005). Idealized, inaccurate but successful: a pragmatic approach to evaluating models in theoretical ecology. *Biology and Philosophy* 20, 231–255.

Olsen, E. M., Heino, M., Lilly, G. R. *et al.* (2004). Maturation trends indicative of rapid evolution preceded the collapse of northern cod. *Nature* 428, 932–935.

Orzack, S. H. and Hines, W. G. S. (2005). The evolution of strategy variation: will an ESS evolve? *Evolution* 59, 1183–1193.

Otto, S. P. and Day, T. (2006). *A Biologist's Guide to Mathematical Modeling.* Princeton: Princeton University Press.

Page, K. M. and Nowak, M. A. (2002). Unifying evolutionary dynamics. *Journal of Theoretical Biology* 219, 93–98.

Parker, G. (2006). Sexual conflict over mating and fertilization: an overview. *Philosophical Transactions of the Royal Society of London, Series B* 361, 235–259.

Parker, G. A. (1979). Sexual selection and sexual conflict. In *Sexual Selection and Reproductive Competition in Insects*, eds. M. S. Blum and N. A. Blum, pp. 123–166. New York: Academic Press.

Parker, G. A. and Maynard Smith, J. (1990). Optimality theory in evolutionary biology. *Nature* 348, 27–33.

Parmigiani, S., Torricelli, P. and Lugli, M. (1987). Intermale aggression in *Padogobius martensi* (Guenther) (Pisces: Gobiidae) during the breeding season: effects of size, prior residence, and parental investment. *Monitore Zoologico Italiano* 22, 161–170.

Parvinen, K. (2004). Adaptive responses to landscape disturbances: theory. In *Evolutionary Conservation Biology*, eds. R. Ferrière, U. Dieckmann and D. Couvet, pp. 265–283. Cambridge: Cambridge University Press.

Peck, S. L. (2004). Simulation as experiment: a philosophical reassessment for biological modeling. *Trends in Ecology and Evolution* 19, 530–534.

Pen, I. (2000). Reproductive effort in viscous populations. *Evolution* 54, 293–297.

Pen, I. and Weissing, F. J. (2000). Towards a unified theory of cooperative breeding: the role of ecology and life history re-examined. *Proceedings of the Royal Society of London, Series B* 267, 2411–2418.

Penn, D. J. (2003). The evolutionary roots of our environmental problems: toward a Darwinian ecology. *Quarterly Review of Biology* 78, 275–301.

Peters, A. D. and Lively, C. M. (1999). The Red Queen and fluctuating epistasis: a population genetic analysis of antagonistic coevolution. *American Naturalist* 154, 393–405.

Piattelli-Palmarini, M. (1994). *Inevitable Illusions: How Mistakes of Reason Rule Our Minds.* Chichester, UK: John Wiley.

Pigliucci, M. (2005). Evolution of phenotypic plasticity: where are we going now? *Trends in Ecology and Evolution* 20, 481–486.

Pigliucci, M. (2006). Genetic variance–covariance matrices: a critique of the evolutionary quantitative genetics research program. *Biology and Philosophy* 21, 1–23.

Pigliucci, M. and Schlichting, C. D. (1997). On the limits of quantitative genetics for the study of phenotypic evolution. *Acta Biotheoretica* 45, 143–160.

Pravosudov, V. V. and Lucas, J. R. (2001). Daily patterns of energy storage in food-caching birds under variable daily predation risk: a dynamic state variable model. *Behavioral Ecology and Sociobiology* 50, 239–250.

Queller, D. C. (1997). Why do females care more than males? *Proceedings of the Royal Society of London, Series B* 264, 1555–1557.

Radwan, J., Unrug, J., Snigórska, K. and Gawronska, K. (2004). Effectiveness of sexual selection in preventing fitness deterioration in bulb mite populations under relaxed natural selection. *Journal of Evolutionary Biology* 17, 94–99.

Reale, D., Bousses, P. and Chapuis, J. L. (1996). Female-biased mortality induced by male sexual harassment in a feral sheep population. *Canadian Journal of Zoology* 74, 1812–1818.

Rhen, T. (2000). Sex-limited mutations and the evolution of sexual dimorphism. *Evolution* 54, 37–43.

Rice, S. H. (2004). *Evolutionary Theory: Mathematical and Conceptual Foundations*. Sunderland: Sinauer.

Rodenhouse, N. L., Sherry, T. W. and Holmes, R. T. (1997). Site-dependent regulation of population size: a new synthesis. *Ecology* 78, 2025–2042.

Roff, D. A. (1994). Optimality modeling and quantitative genetics: a comparison of the two approaches. In *Quantitative Genetic Studies of Behavioral Evolution*, ed. C. R. B. Boake, pp. 49–66. Chicago: Chicago University Press.

Roff, D. A. (1997). *Evolutionary Quantitative Genetics*. New York: Chapman & Hall.

Roff, D. A. (2000). Trade-offs between growth and reproduction: an analysis of the quantitative genetic evidence. *Journal of Evolutionary Biology* 13, 434–445.

Ronce, O. and Kirkpatrick, M. (2001). When sources become sinks: migrational meltdown in heterogeneous habitats. *Evolution* 55, 1520–1531.

Rosales, A. (2005). John Maynard Smith and the natural philosophy of adaptation. *Biology and Philosophy* 20, 1027–1040.

Roth, T. C. II., Lima, S. L. and Vetter, W. E. (2006). Determinants of predation risk in small wintering birds: the hawk's perspective. *Behavioral Ecology and Sociobiology* 60, 195–204.

Rousset, F. (2004). *Genetic Structure and Selection in Subdivided Populations*. Princeton: Princeton University Press.

Rowe, L., Ludwig, D. and Schluter, D. (1994). Time, condition and the seasonal decline of avian clutch size. *American Naturalist* 143, 698–722.

Roze, D. and Rousset, F. (2005). Inbreeding depression and the evolution of dispersal rates: a multilocus model. *American Naturalist* 166, 708–721.

Ryan, M. J. (1998). Sexual selection, receiver biases, and the evolution of sex differences. *Science* 281, 1999–2003.

Sachs, J. L., Mueller, U. G., Wilcox, T. P. and Bull, J. J. (2004). The evolution of cooperation. *Quarterly Review of Biology* 79, 135–160.

Sarkar, S. (2005). Maynard Smith, optimization, and evolution. *Biology and Philosophy* 20, 951–966.

Sasaki, A. and de Jong, G. (1999). Density dependence and unpredictable selection in a heterogeneous environment: compromise and polymorphism in the ESS reaction norm. *Evolution* 53, 1329–1342.

Schlichting, C. D. and Pigliucci, M. (1998). *Phenotypic Evolution: A Reaction Norm Perspective*. Sunderland: Sinauer.

Servedio, M. R. and Lande, R. (2006). Population genetic models of male and mutual mate choice. *Evolution* 60, 674–685.

Shine, R., LeMaster, M. P., Moore, I. T., Olsson, M. M. and Mason, R. T. (2001). Bumpus in the snake den: effects of sex, size, and body condition on mortality of red-sided garter snakes. *Evolution* 55, 598–604.

Shulman, M. J. (1990). Aggression among sea urchins on Caribbean coral reefs. *Journal of Experimental Marine Biology and Ecology* 140, 197–207.

Shuster, S. M. and Wade, M. J. (2003). *Mating Systems and Strategies*. Princeton: Princeton University Press.

Stephens, P. A. and Sutherland, W. J. (1999). Consequences of the Allee effect for behaviour, ecology and conservation. *Trends in Ecology and Evolution* 10, 401–405.

Steppan, S. J., Phillips, P. C. and Houle, D. (2002). Comparative quantitative genetics: evolution of the G matrix. *Trends in Ecology and Evolution* 17, 320–327.

Svensson, E. I., Abbott, J. and Härdling, R. (2005). Female polymorphism, frequency dependence, and rapid evolutionary dynamics in natural populations. *American Naturalist* 165, 567–576.

Tamachi, N. (1987). The evolution of alarm calls: an altruism with nonlinear effect. *Journal of Theoretical Biology* 127, 141–153.

Tanaka, Y. (1998). Theoretical aspects of extinction by inbreeding depression. *Researches on Population Ecology* 40, 279–286.

Taylor, P. D. (1996). Inclusive fitness arguments in genetic models of behaviour. *Journal of Mathematical Biology* 34, 654–674.

Taylor, P. and Frank, S. (1996). How to make a kin selection model. *Journal of Theoretical Biology* 180, 27–37.

Taylor, P. D., Wild, G. and Gardner, A. (2007). Direct fitness or inclusive fitness: how shall we model kin selection? *Journal of Evolutionary Biology* 20, 301–309.

Trayhum, P. (1993). Brown adipose tissue: from thermal physiology to bioenergetics. *Journal of Bioscience* 18, 161–173.

Tregenza, T. (1997). Darwin a better name than Wallace? *Nature* 385, 480.

van Doorn, G. S., Dieckmann, U. and Weissing, F. J. (2004). Sympatric speciation by sexual selection: a critical reevaluation. *American Naturalist* 163, 709–725.

Visser, M. E., Both, C. and Lambrechts, M. M. (2004). Global climate change leads to mistimed avian reproduction. *Advances in Ecological Research* 35, 89–110.

Wade, M. J., Shuster, S. M. and Demuth, J. P. (2003). Sexual selection favors female-biased sex ratios: the balance between the opposing forces of sex-ratio selection and sexual selection. *American Naturalist* 162, 403–414.

Waxman, D. and Gavrilets, S. (2005). 20 questions on adaptive dynamics. *Journal of Evolutionary Biology* 18, 1139–1154.

Weber, T. P. and Houston, A. I. (1997). A general model for time-minimising avian migration. *Journal of Theoretical Biology* 185, 447–458.

Weinig, C., Johnston, J., German, Z. M. and Demink, L. M. (2006). Local and global costs of adaptive plasticity to density in *Arabidopsis thaliana*. *American Naturalist* 167, 826–836.

Weissing, F. J. (1996). Genetic versus phenotypic models of selection: can genetics be neglected in a long-term perspective? *Journal of Mathematical Biology* 34: 533–555.

Wenseleers, T. (2006). Modelling social evolution: the relative merits and limitations of a Hamilton's rule-based approach. *Journal of Evolutionary Biology* 19, 1419–1422.

Wilkins, J. F. and Haig, D. (2003). What good is genomic imprinting: the function of parent-specific gene expression. *Nature Reviews Genetics* 4, 359–368.

Williams, C. G. (1985). A defence of reductionism in evolutionary biology. *Oxford Surveys in Evolutionary Biology* 2, 1–27.

Witter, M. S., Cuthill, I. C. and Bonser, R. H. C. (1994). Experimental investigations of mass-dependent predation risk in the European starling, *Sturnus vulgaris*. *Animal Behaviour* 48, 201–222.

Wolf, J. B. and Wade, M. J. (2001). On the assignment of fitness to parent and offspring: whose fitness is it and when does it matter? *Journal of Evolutionary Biology* 14, 347–356.

Index

Printed in the United States
by Baker & Taylor Publisher Services